INVESTING IN FEDERAL FACILITIES

Asset Management Strategies for the 21st Century

Committee on Business Strategies for Public Capital Investment

Board on Infrastructure and the Constructed Environment

Division on Engineering and Physical Sciences

NATIONAL RESEARCH COUNCIL
OF THE NATIONAL ACADEMIES

THE NATIONAL ACADEMIES PRESS
Washington, D.C.
www.nap.edu

THE NATIONAL ACADEMIES PRESS 500 Fifth Street, N.W. Washington, DC 20001

NOTICE: The project that is the subject of this report was approved by the Governing Board of the National Research Council, whose members are drawn from the councils of the National Academy of Sciences, the National Academy of Engineering, and the Institute of Medicine. The members of the committee responsible for the report were chosen for their special competences and with regard for appropriate balance.

This study was supported by Contract No. SLMAQM00C6017 between the National Academy of Sciences and the Department of State on behalf of the Federal Facilities Council. Any opinions, findings, conclusions, or recommendations expressed in this publication are those of the authors and do not necessarily reflect the views of the organizations or agencies that provided support for this project.

International Standard Book Number 0-309-08919-0 (Book)
International Standard Book Number 0-309-50857-6 (PDF)

Available from:

Board on Infrastructure and the Constructed Environment
National Research Council
500 Fifth Street, N.W.
Washington, DC 20001

Additional copies of this report are available from the National Academies Press, 500 Fifth Street, N.W., Lockbox 285, Washington, DC 20055; (800) 624-6242 or (202) 334-3313 (in the Washington metropolitan area); Internet, http://www.nap.edu

Copyright 2004 by the National Academy of Sciences. All rights reserved.

Printed in the United States of America

THE NATIONAL ACADEMIES
Advisers to the Nation on Science, Engineering, and Medicine

The **National Academy of Sciences** is a private, nonprofit, self-perpetuating society of distinguished scholars engaged in scientific and engineering research, dedicated to the furtherance of science and technology and to their use for the general welfare. Upon the authority of the charter granted to it by the Congress in 1863, the Academy has a mandate that requires it to advise the federal government on scientific and technical matters. Dr. Bruce M. Alberts is president of the National Academy of Sciences.

The **National Academy of Engineering** was established in 1964, under the charter of the National Academy of Sciences, as a parallel organization of outstanding engineers. It is autonomous in its administration and in the selection of its members, sharing with the National Academy of Sciences the responsibility for advising the federal government. The National Academy of Engineering also sponsors engineering programs aimed at meeting national needs, encourages education and research, and recognizes the superior achievements of engineers. Dr. Wm. A. Wulf is president of the National Academy of Engineering.

The **Institute of Medicine** was established in 1970 by the National Academy of Sciences to secure the services of eminent members of appropriate professions in the examination of policy matters pertaining to the health of the public. The Institute acts under the responsibility given to the National Academy of Sciences by its congressional charter to be an adviser to the federal government and, upon its own initiative, to identify issues of medical care, research, and education. Dr. Harvey V. Fineberg is president of the Institute of Medicine.

The **National Research Council** was organized by the National Academy of Sciences in 1916 to associate the broad community of science and technology with the Academy's purposes of furthering knowledge and advising the federal government. Functioning in accordance with general policies determined by the Academy, the Council has become the principal operating agency of both the National Academy of Sciences and the National Academy of Engineering in providing services to the government, the public, and the scientific and engineering communities. The Council is administered jointly by both Academies and the Institute of Medicine. Dr. Bruce M. Alberts and Dr. Wm. A. Wulf are chair and vice chair, respectively, of the National Research Council.

www.national-academies.org

COMMITTEE ON BUSINESS STRATEGIES FOR PUBLIC CAPITAL INVESTMENT

ALBERT A. DORMAN, NAE, *Chair*, AECOM, Los Angeles
DAVID NASH, RADM, CEC USN (retired), *Vice Chair*, BE & K, Birmingham, Alabama
ADJO AMEKUDZI, Georgia Institute of Technology, Atlanta
KIMBALL J. BEASLEY, Wiss, Janey, Elstner Associates, Inc., New York
JEFFERY CAMPBELL, Brigham Young University, Provo, Utah
ERIC T. DILLINGER, Carter and Burgess, Inc., Fort Worth, Texas
JAMES R. FOUNTAIN, JR., Governmental Accounting Standards Board, Norwalk, Connecticut
THOMAS K. FRIDSTEIN, Hillier, New York
LUCIA E. GARSYS, Quality Services Officer, Hillsborough County, Florida
DAVID L. HAWK, New Jersey Institute of Technology, Newark
RALPH L. KEENEY, NAE, Duke University, Durham, North Carolina
STEPHEN J. LUKASIK, Independent Consultant, Los Angeles
CAROL Ó'CLÉIREACÁIN, Brookings Institution and Independent Consultant, New York
CHARLES SPRUILL, Fannie Mae, Washington, D.C.

Staff

LYNDA STANLEY, Study Director
RICHARD LITTLE, Director, Board on Infrastructure and the Constructed Environment
CAMERON GORDON, Program Officer
JASON DREISBACH, Research Associate
DANA CAINES, Financial Associate
PAT WILLIAMS, Senior Project Assistant

BOARD ON INFRASTRUCTURE AND THE CONSTRUCTED ENVIRONMENT

PAUL GILBERT, *Chair*, Parsons, Brinckerhoff, Quade, and Douglas, Seattle
MASSOUD AMIN, University of Minnesota, Minneapolis
RACHEL DAVIDSON, Cornell University, Ithaca, New York
REGINALD DESROCHES, Georgia Institute of Technology, Atlanta
DENNIS DUNNE, California Department of General Services, Sacramento
PAUL FISETTE, University of Massachusetts, Amherst
WILLIAM H. HANSMIRE, Parsons, Brinckerhoff, Quade, and Douglas, San Francisco
HENRY HATCH, U.S. Army Corps of Engineers (retired), Oakton, Virginia
AMY HELLING, Georgia State University, Atlanta
SUE McNEIL, University of Illinois, Chicago
DEREK PARKER, Anshen+Allen, San Francisco
DOUGLAS SARNO, The Perspectives Group, Inc., Alexandria, Virginia
HENRY G. SCHWARTZ, JR., Washington University, St. Louis
DAVID SKIVEN, General Motors Corporation, Detroit
MICHAEL STEGMAN, University of North Carolina, Chapel Hill
WILLIAM WALLACE, Rensselaer Polytechnic Institute, Troy, New York
ZOFIA ZAGER, Fairfax County, Virginia
CRAIG ZIMRING, Georgia Institute of Technology, Atlanta

Staff

RICHARD LITTLE, Director, Board on Infrastructure and the Constructed Environment
LYNDA STANLEY, Executive Director, Federal Facilities Council
MICHAEL COHN, Program Officer
DANA CAINES, Financial Associate
PAT WILLIAMS, Senior Project Assistant

Chairman's Foreword

Many segments of government have come to believe that an opportunity exists to introduce more objectivity into the politically sensitive issues and processes surrounding public investment in federal facilities. The U.S. General Accounting Office's designation of federal real property as a government-wide high-risk area on January 30, 2003, now makes it urgent to seize the opportunity. This committee, while recognizing the daunting complexities of the challenge, has nonetheless attempted to indicate some directions such a quest might take.

In accordance with its designation as the Committee on Business Strategies for Public Capital Investment, the committee reviewed principles, policies, and practices used by a range of private-sector organizations ("businesses") in making decisions about facilities investments. The committee recognized early on that government and for-profit organizations have inherently different missions and service orientations and different ways of operating, making decisions, and measuring success. Within government, the same types of differences exist among departments and agencies. *The committee concluded that there is no single solution from the private sector that could apply to all federal facilities investment and management, nor should we expect that one will be found. Nevertheless, there are private-sector principles, policies, and practices integral to successful facilities investment and management decisions that appear suitable for conversion into equivalent federal precepts.* This report enumerates these precepts, elaborates on them, and suggests techniques for adapting them to the federal operating environment.

Just as there is no panacea for federal facilities investment and management, there is no substitute for good decision makers. Decision theories and processes, criteria, rules and regulations, no matter how advanced, are only tools. The federal operating environment is a complex system of differing value judgments, a wide array of justifiable goals and objectives, changing missions, interlocking authorities and responsibilities, and legitimate constituency pressures that must always be balanced against the resources judged available. Therefore, the committee also emphasizes the human resources aspects: the development of good decision makers at all levels and the creation of an atmosphere of mutual respect and trust between them.

In further recognition of this complex environment, the committee has outlined an implementation program that suggests how elected officials and the many dedicated and competent career public servants might together develop legislation and guidelines to improve public investment in federal facilities. The effect on the economy of properly directing the billions of dollars expended annually for federal facilities, coupled with recognition of the impact that these facilities investments have on shaping the environment of 280 million Americans, mandates an early, continuous, and collaborative effort to transform current decision-making processes.

Albert A. Dorman
Chair, Committee on Business Strategies for Public Capital Investment

Preface

At a fundamental level, choices made today about investments in facilities and infrastructure[1] directly affect the future quality of shelter, workplaces, and the delivery of services. When, where, and how to invest in facilities are critical variables for determining that quality.

During the past 20 years, numerous studies have focused on the deteriorating condition of infrastructure throughout the United States, including the deteriorating condition of facilities owned and leased by the federal government. Over the same period, the operating environments of both private and public-sector organizations have been evolving in response to rapid advances in technologies, changes in demographics, and increasingly rapid changes in society at large. These changes both require and make possible new approaches to facilities and infrastructure investment and management.

Under successive administrations, there has been a concerted effort to make the federal government more responsive to its citizens, more accountable for what it does, more performance- and results-oriented, and more open to innovative approaches, with all of these attributes being seen as "businesslike." Elected officials, senior agency executives, and facilities managers have asked, Can the experience of private-sector organizations with facilities investment and management provide insight for similar decisions and responsibilities facing the federal government?

---

[1] In this report, facilities investments are defined as new construction, renewal, maintenance, retrofitting, acquiring, leasing, and decommissioning or disposing of buildings, structures, and their supporting infrastructure.

STUDY APPROACH

The sponsoring agencies of the Federal Facilities Council (FFC)[2] formulated the request for the current study with these questions in mind. In 2002, the National Research Council (NRC) appointed the Committee on Business Strategies for Public Capital Investment to undertake the following task:

> Develop guidelines for making improved public investment decisions about facilities and supporting infrastructure, their maintenance, renewal, replacement, and decommissioning. As part of this task, the committee was asked to review and appraise current practices used to support facilities decision-making in both the private and public sectors and identify objectives, practices, and performance measures to help determine appropriate levels of investment.

In discharging its task, the committee recognized at the outset that there are inherent differences in the missions, goals, and operating environments of private-sector organizations and those of the federal government, and it elaborates on these and other differences throughout this report. Nonetheless, there are also many similarities in regard to facilities investments. Large organizations of any type must answer two different but related questions: What facilities are needed to support the organization's mission? How should decisions about facilities investments be made if organizational goals and objectives are to be met?

The 14 committee members have expertise in the operation and management of large private and public-sector organizations, capital investment, facilities programming and management, corporate real estate, building performance and serviceability, government budgeting and finance, decision sciences, economics, and architecture and engineering. In addition, many of the committee members are involved in professional organizations that focus on facilities-related issues, including the American Institute of Architects, the American Planning Association, the American Society of Civil Engineers, the Society of American Military Engineers, the Association of Higher Education Facilities Officers, the International Facility Management Association, the National Society of Professional Engineers, and the Transportation Research Board. Biographical information about the committee members is provided in Appendix A.

As one of its research activities, the committee interviewed representatives of private-sector corporations, federal agencies, other public entities, and not-for-profit organizations who are responsible for facilities investment decisions. Persons interviewed and their affiliations and other persons who provided information to the committee are listed in Appendix B. Appendix C contains the interview discussion outline.

---

[2]The FFC is a cooperative association of 24 federal departments and agencies operating under the aegis of the National Research Council. The FFC's mission is to identify and advance technologies, processes, and management practices that improve the performance of federal facilities over their entire life cycle, from planning to disposal.

PREFACE   xi

During 22 months of committee, subcommittee, individual, and staff work and five deliberative 2-day meetings, the committee also independently collected, studied, analyzed, and compared federal, other public, private, and not-for-profit sector facilities investment and management principles, policies, and practices. Based on this research and on their individual and collective experience, the committee identified a set of principles and policies that it believes are highly effective and could be beneficially adapted for use within the federal government.

CONTENTS OF THE REPORT

This study reviews how decisions for private- and public-sector facilities investments are being made in today's operating environments and the roles of the various groups and individuals who make the decisions. The intent of the committee is to provide specific recommendations to improve decision-making and management processes so that the resources available for federal facilities investments can be allocated more effectively and the results can be measured. To this end, the study addresses such questions as, How can the various parties to federal facilities decisions be motivated to act in the public's long-term interest, given short-term election cycles and budgets and the recognition that the results of decisions made today may not be apparent for many years? Are there better methods to align federal departments' and agencies' portfolios of facilities with their missions? Can the climate for making investment decisions about federal facilities be improved? When should federal facilities be owned or leased or disposed of?

This report is addressed to a wide audience: decision makers in Congress, federal departments, agencies, and their advisors; federal facilities program managers, operating groups, and their contractors; and program and budget analysts throughout the federal government. Decision makers, facilities program managers, and program and budget analysts in public agencies at the state and local levels may also find value in the report since they face many of the same issues as their federal counterparts. Because this report addresses multiple audiences, different readers will find different chapters to be of greatest interest. For those with limited time, the Executive Summary and Chapter 6 should be read together.

Chapter 1, "Context," quantifies the ongoing investment in federal facilities, identifies some fundamental characteristics of the private sector and the federal government that affect facilities investments, looks at the dynamic nature of facilities requirements as compared with the longevity and life cycles of facilities, and discusses some conceptual shifts in facilities investment decision making.[3]

---

[3]In this and other chapters, a number of sources are cited in regard to the value of facilities and the level of investments in facilities made by public and private-sector organizations. No attempt has been made to reconcile the numbers across the various sources. For this report, the numbers are primarily cited to convey the magnitude of the investments involved.

Chapter 2, "Facilities Asset Management," describes how facilities management practices are evolving to better support organizational objectives and decision making and to better manage portfolios of facilities, as well as the additional skills that are now required of facilities asset managers. Chapter 3, "Decision Making to Support Organizational Missions," describes how best-practice organizations use their mission as guidance for facilities investment decisions; why and how they create frameworks for facilities investment decision making and management; basic issues related to facilities investments; and decision-making processes. Chapter 4, "Environments for Effective Decision Making," focuses on how best-practice organizations foster open communications and build trust among the various stakeholder groups to create an environment for effective decision making. The use of performance measures, continuous feedback procedures, accountability, and incentives to evaluate and improve the outcomes of decision-making processes are featured. Chapter 5, "Alternative Approaches for Acquiring Federal Facilities," describes public-private partnerships and a range of other approaches that could be tested more widely to leverage funding for federal facilities investments. Chapter 6, "Adapting Principles and Policies from Best-Practice Organizations to the Federal Operating Environment," reviews issues and obstacles when adapting principles and policies from best-practice organizations for use in the federal operating environment. The committee sets forth 15 recommendations for adapting and implementing these principles and policies and concludes by offering an overall strategy for their implementation.

TERMS USED IN THIS REPORT

Terminology varies across the fields of facilities management, finance, budgeting, accounting, and economics. For example, terms like "capital" are used in all of these fields but defined differently. This can sometimes lead to confusion and miscommunication when engineers, financial and budget analysts, accountants, economists, and elected officials work together. In an effort to clearly communicate the committee's intent, key terms used in this report are explained below. Where the committee has used a definition from another source, the source is cited.

Best-practice organizations. Private-sector, not-for-profit, and public organizations that use principles, policies, and practices that the committee—through its research, interviews, collective and individual experience, and systematic

---

Similarly, there were many sources of data on the amount of facility space owned and leased by the federal government and the types of space. Again, the numbers are cited to convey the magnitude and diversity of the federal government's holdings, with no attempt to reconcile data differences across the sources.

analysis—has determined to be highly effective for facilities investment decision making and asset management.

Business case analysis. Tool for planning and decision making that projects the financial implications and other organizational consequences of a proposed action (Schmidt, 2003b). A business case analysis is used to ensure that the objectives for a proposed facility-related investment are clearly defined; that a broad range of alternatives for meeting the objectives is developed; that the alternatives are evaluated to determine how well the objectives will be met; and that tradeoffs are explicit. It is a living tool that is continually revisited, refined, and updated. Although at its heart the business case is a financial analysis, it also contains information on organizational impacts that cannot be quantified in monetary terms, such as mission-readiness or fulfillment, customer satisfaction, and public image.

Facilities asset management. Systematic process of maintaining, upgrading, and operating physical assets cost-effectively. It combines engineering principles with sound business practices and economic theory and provides tools to achieve a more organized, logical approach to decision making (FHWA, 1999). A facilities asset management approach allows for both program- or network-level management and project-level management and thereby supports both executive-level (portfolio of facilities) and field-level decision making.

Facilities investments. New construction, renewal, maintenance, retrofitting, acquiring, leasing, and decommissioning or disposing of buildings, structures, and their supporting infrastructure. Investments in land are excluded.

Not-for-profit organizations. Groups organized for purposes other than generating profit and in which no part of the organization's net earnings may inure to the benefit of any private shareholder or individual. Not-for-profit organizations may take many forms, including that of a corporation, an individual enterprise, an unincorporated association, a partnership, or a charitable foundation. They must be designated as not-for-profit at their inception and are governed by state laws.

Private-sector organizations. Enterprises formed to engage in activities that generate profit for their owners or shareholders. They can take a number of forms and legal definitions—sole proprietorships, general partnerships, limited partnerships, joint ventures, C corporations, limited liability corporations, and S corporations, among others.

Pro forma statement. Strictly financial analysis included in a business case analysis.

Acknowledgments of Committee Members and Reviewers

The Committee on Business Strategies for Public Capital Investment acknowledges and thanks all those representatives of private-sector organizations, federal agencies, other public entities, and not-for-profit institutions who provided background information and shared their personal expertise through briefings and interviews.

The chair of the committee expresses his personal appreciation to all of the committee members for sharing their expertise, views, and opinions; for making substantial contributions to concepts and text; and for giving generously of their time.

This report has been reviewed in draft form by individuals chosen for their diverse perspectives and technical expertise, in accordance with procedures approved by the National Research Council's Report Review Committee. The purpose of this independent review is to provide candid and critical comments that will assist the institution in making its published report as sound as possible and to ensure that the report meets institutional standards for objectivity, evidence, and responsiveness to the study charge. The review comments and draft manuscript remain confidential to protect the integrity of the deliberative process. We wish to thank the following individuals for their review of this report:

David G. Cotts, author and management consultant,
Dennis D. Dunne, California Department of General Services (retired),
Carl Ference, Trammell Crow Company,
Amy Helling, Georgia State University,
James C. Hershauer, Arizona State University,
Steven Kelman, Harvard University, and
Morris Tanenbaum, AT&T Corporation (retired).

Although the reviewers listed above have provided many constructive comments and suggestions, they were not asked to endorse the conclusions or recommendations, nor did they see the final draft of the report before its release. The review of this report was overseen by Dale F. Stein, President Emeritus, Michigan Technological University. Appointed by the National Research Council, he was responsible for making certain that an independent examination of this report was carried out in accordance with institutional procedures and that all review comments were carefully considered. Responsibility for the final content of this report rests entirely with the authoring committee and the institution.

Contents

EXECUTIVE SUMMARY	1
1 CONTEXT	13

   Background, 13
   The Ongoing Investment in Federal Facilities, 14
   Some Characteristics of Private-Sector Organizations That Affect
      Facilities Investment and Management, 16
   Some Characteristics of the Federal Government That Affect Facilities
      Investment and Management, 20
   Facilities Requirements, Longevity, and Life-Cycle Costs, 25
   Conceptual Shifts in Facilities Investment Decision Making, 28

2 FACILITIES ASSET MANAGEMENT	30

   Background, 30
   Facilities Asset Management, 32
   Components of a Facilities Asset Management Approach, 32
   Facilities Asset Managers, 37
   Examples of Facilities Asset Management Systems, 40
   Principles and Policies from Best-Practice Organizations, 43

3 DECISION MAKING TO SUPPORT ORGANIZATIONAL MISSIONS	44

   Background, 44
   The Roles of Analysis and Values in Decision Making, 45
   Management Approaches for Achieving a Mission, 47

Information for Decision Making, 50
Decision-Making Processes, 55
Principles and Policies from Best-Practice Organizations, 59

4 ENVIRONMENTS FOR EFFECTIVE DECISION MAKING   62
Background, 62
Open Communications, Trust, and Credible Information, 62
Performance Measures, 65
Evaluations and Continuous Feedback, 68
Forms of Feedback, 69
Accountability, 72
Incentives, 73
Principles and Policies from Best-Practice Organizations, 73

5 ALTERNATIVE APPROACHES FOR ACQUIRING   76
FEDERAL FACILITIES
Background, 76
Issues Related to Full Up-front Funding of Facilities, 77
Issues Related to the Use of Alternative Approaches for
    Acquiring Facilities, 78
Summary and a Recommendation, 88

6 ADAPTING PRINCIPLES AND POLICIES FROM   89
BEST-PRACTICE ORGANIZATIONS TO THE
FEDERAL OPERATING ENVIRONMENT
Background, 89
Special Aspects of the Federal Operating Environment, 90
Adapting Best-Practice Principles and Policies to the
    Federal Environment, 93
An Overall Strategy for Implementation, 113

BIBLIOGRAPHY   118

APPENDIXES

A Biographical Sketches of Committee Members   127
B Committee Interviews and Briefings   134
C Interview Discussion Outline   137

List of Figures and Tables

FIGURES

1.1 Federal agencies' facilities holdings in millions of square feet, 15
1.2 Distribution of federal government space by type of use, 15
1.3 Distribution of total assets for a typical corporate organization, 17
1.4 The various stakeholders in facilities investments and their diverse and overlapping objectives, 22
1.5 Facility life cycle, 27

2.1 The evolving focus of facilities asset management, 30
2.2 Factors driving the evolution of facilities management, 31
2.3 Components of a facilities asset management system, 33
2.4 Linking organizational goals with facilities investment and operations, 33
2.5 A facilities asset management structure (BYU), 41
2.6 A facilities asset management framework (BYU), 42

3.1 Typical decision-making process for facilities investments, 56

6.1 A sociotechnical system view for decision making, 114
6.2 A model for integrating scientific and social values in decision making, 115

TABLES

2.1 Skills Required by Facilities Asset Managers, 38
2.2 Business Skills for the Facility Manager, 39

3.1 An Approach for Nonmanufacturing Facilities (GM), 50

4.1 Strategic Assessment Model Matrix of the Association of Higher Education Facilities Officers (APPA), 67

Executive Summary

Federal facilities investments are matters of public policy. The facilities acquired by the federal government provide a means to produce and distribute public goods and services to 280 million Americans, create jobs, strengthen the national economy, and support the missions of federal departments and agencies, including the defense and security missions. Such investments also support policies for public transportation, urban revitalization, and historic preservation, among others.

Hundreds of billions of dollars have been invested in federal facilities and their associated infrastructure. As of September 2000, the federal government owned or leased 3.3 billion square feet of space worldwide, distributed across more than 500,000 facilities, conservatively valued at $328 billion. Annually, it spends upwards of $21 billion for the acquisition and renovation of facilities, approximately $4.5 billion to power, heat, and cool its buildings, and more than $500 million for water and waste disposal. Additional expenditures for facilities maintenance, repair, renewal, demolition, and security upgrades probably amount to billions of dollars per year but are not readily identifiable under the current budget structure.

Despite the magnitude of this ongoing investment, federal facilities continue to deteriorate, backlogs of deferred maintenance continue to increase, and excess, underutilized, and obsolete facilities continue to consume limited resources. Many departments and agencies have the wrong facilities, too many or not enough facilities, or facilities that are poorly sited to support their missions. Such facilities constitute a drain on the federal budget in actual costs and in foregone opportunities to invest in other public resources and programs.

On January 30, 2003, the U.S. General Accounting Office (GAO) designated federal real property as a government-wide high-risk area[1] because current trends "have multibillion dollar cost implications and can seriously jeopardize mission accomplishment" and because "federal agencies face many challenges securing real property due to the threat of terrorism." It declared that "current structures and processes may not be adequate to address the problems," so that "a comprehensive, integrated transformation strategy" may be required.

PRINCIPLES AND POLICIES FOR FACILITIES INVESTMENTS AND MANAGEMENT

As the committee reviewed the types of analyses, the processes, and the decision-making environments that private-sector and other organizations use for facilities investments and management, it focused on identifying principles and policies used by best-practice organizations, as defined by the committee. The committee found that, in matters of facilities investment and management, best-practice organizations do the following:

Principle/Policy 1. Establish a framework of procedures, required information, and valuation criteria that aligns the goals, objectives, and values of their individual decision-making and operating groups to achieve the organization's overall mission; create an effective decision-making environment; and provide a basis for measuring and improving the outcomes of facilities investments. The components of the framework are understood and used by all leadership and management levels.

Principle/Policy 2. Implement a systematic facilities asset management approach that allows for a broad-based understanding of the condition and functionality of their facilities portfolios—as distinct from their individual projects—in relation to their organizational missions. Best-practice organizations ensure that their facilities and infrastructure managers possess both the technical expertise and the financial analysis skills to implement a portfolio-based approach.

Principle/Policy 3. Integrate facilities investment decisions into their organizational strategic planning processes. Best-practice organizations evaluate facilities investment proposals as mission enablers rather than solely as costs.

---

[1] GAO's high-risk update is provided at the start of each new Congress. The reports are intended to help the new Congress "focus its attention on the most important issues and challenges facing the federal government." (GAO, 2003f)

Principle/Policy 4. Use business case analyses to rigorously evaluate major facilities investment proposals and to make transparent a proposal's underlying assumptions; the alternatives considered; a full range of costs and benefits; and the potential consequences for their organizations.

Principle/Policy 5. Analyze the life-cycle costs of proposed facilities, the life-cycle costs of staffing and equipment inherent to the proposal, and the life-cycle costs of the required funding.

Principle/Policy 6. Evaluate ways to disengage from, or exit, facilities investments as part of the business case analysis and include disposal costs in the facilities life-cycle cost to help select the best solution to meet the requirement.

Principle/Policy 7. Base decisions to own or lease facilities on the level of control required and the planning horizon for the function, which may or may not be the same as the life of the facility.

Principle/Policy 8. Use performance measures in conjunction with both periodic and continuous long-term feedback to evaluate the results of facilities investments and to improve the decision-making process itself.

Principle/Policy 9. Link accountability, responsibility, and authority when making and implementing facilities investment decisions.

Principle/Policy 10. Motivate employees as individuals and as groups to meet or exceed accepted levels of performance by establishing incentives that encourage effective decision making and reward extraordinary performance.

ADAPTING THE PRINCIPLES AND POLICIES TO THE FEDERAL OPERATING ENVIRONMENT

Adapting the aforementioned principles and policies for facilities investments for use by the federal government requires consideration of and compensation for a number of special aspects of the federal operating environment. These aspects include the goals and missions of the federal government, its departments, and agencies; the organizational structure and decision-making environment; the nature of federal facilities investments; and the annual budget process and its attendant procedures. They are described more fully in Chapters 1 and 6.

Despite the inherent differences, the committee's overall conclusion is that aspects of *all* of the identified principles and policies used by best-practice organizations *can* be adapted in varying form to the federal operating environment. It has therefore made recommendations to aid in developing an overall framework based on suitable adaptations of the identified principles and policies.

The committee also concluded that there is no single solution from the private sector that can be applied to all issues related to federal facilities investment and management, nor should there be an expectation that one will be found. The committee points to the number of missions and the variation in size, resources, culture, and political support of the many federal agencies with facilities-related responsibilities and urges all involved not to attempt to create one-size-fits-all solutions to different problems.

Instead, the committee recommends that efforts be made to concurrently and collaboratively develop top-down and bottom-up approaches while keeping in mind differences among various agency missions and cultures as well as similarities in many specifics of facility investment and management. Varying practices within common principles and policies should be expected.

RECOMMENDATION 1. The federal government should adopt a framework of procedures, required information, and valuation criteria for federal facilities investment decision making and management that incorporates all of the principles and policies enumerated by this committee.

Implementation of a framework that incorporates the identified principles and policies will align the goals, objectives, and values of individual federal decision-making and operating groups with overall missions; create an effective decision-making environment; and provide a basis for measuring and improving the outcomes of federal facilities investments. Because such a framework represents a significant departure from current operating procedures, it may be advisable to establish one or more pilot projects. A small government agency with a diverse portfolio of facilities might provide the environment in which to test the application of the committee's recommendations.

RECOMMENDATION 2(a). Each federal department and agency should update its facilities asset management program to enable it to make investment and management decisions about individual projects relative to its entire portfolio of facilities.

Federal departments and agencies have begun implementing facilities asset management approaches that allow for a broad-based understanding of the condition and functionality of their facilities portfolios. An updated approach should incorporate life-cycle decision making that accounts for all the inherent operating costs (i.e., facilities, staffing, equipment, and information technologies); accurate databases; condition assessments; performance measures; feedback processes; and appropriately adapted business practices.

RECOMMENDATION 2(b). Each federal department and agency should ensure it has the requisite technical and business skills to implement a facilities asset management approach by providing specialized training for its incumbent facilities asset management staff and by recruiting individuals with these skills.

Most federal departments and agencies currently have staff with the requisite technical skills to implement asset management approaches. Less likely to be found are facilities management staff also versed in financial theory, practices, and management. Departments and agencies should provide their incumbent facilities asset management staff with training in business concepts such as financial theory and analysis. Training can be in the form of coursework, seminars, rotational assignments, and other appropriate methods. As job vacancies occur in facilities management operating groups, departments and agencies should seek to recruit and hire staff with the requisite technical and business skills.

RECOMMENDATION 2(c). To facilitate the alignment of each department's and agency's existing facilities portfolios with its missions, Congress and the administration should jointly lead an effort to consolidate and streamline government-wide policies, regulations, and processes related to facilities disposal, which would encourage routine disposal of excess facilities in a timely manner.

Eighty-one separate policies applicable to the disposal of federal facilities have been identified. These include agency-specific legislative mandates and directives and government-wide socioeconomic and environmental policies. The number of policies related to facilities disposal hinders government-wide efforts to expeditiously dispose of unneeded facilities in response to changing requirements.

RECOMMENDATION 2(d). For departments and agencies with many more facilities than are needed for their missions—the Departments of Defense, Energy, State, and Veterans Affairs, the General Services Administration, and possibly others—Congress and the administration should jointly consider implementing extraordinary measures like the process used for military base realignment and closure (BRAC), modified as required to reflect actual experience with BRAC.

Federal agencies are incurring significant costs by operating and maintaining facilities they no longer need to support today's missions. The Department of Defense (DoD) alone estimates it spends $3 to $4 billion each year maintaining excess facilities. The lack of alignment between a department's or agency's mission and its facilities portfolio, coupled with the cost of operating and maintaining excess facilities, may require extraordinary measures to effect improvement, such as the BRAC process used for closing DoD facilities. The government as a whole and the DoD in particular have 15 years of experience and lessons from BRAC. Such lessons can be used to make adjustments to the process to improve it and adapt it to other departments and agencies, as appropriate.

RECOMMENDATION 3. Each federal department and agency should use its organizational mission as guidance for facilities investment decisions and should then integrate facilities investments into its organizational strategic planning processes. Facilities investments should be evaluated as mission enablers, not solely as costs.

Organizational strategic planning that does not include facilities considerations up front fails to account for a potentially substantial portion of the total cost of a program or initiative. Integrating facilities considerations into evaluations of strategic planning alternatives will provide decision makers with better information about the total long-term costs, considerations, and consequences of a particular course of action. To this end, the senior facilities program manager for a department or agency should be directly and continuously involved in the organization's strategic planning processes. This person should be responsible for providing the translation between the agency's mission and its physical assets; identifying alternatives for meeting the mission; identifying the costs, benefits, and potential consequences of the alternatives; and suggesting facilities investments that will reduce overall—that is, portfolio—costs.

RECOMMENDATION 4(a). Each federal department and agency should develop and use a business case analysis for all significant facilities investment proposals to make clear the underlying assumptions, the alternatives considered, the full range of costs and benefits, and potential consequences for the organization and its missions.

There is no standard format for a business case analysis that can be readily adapted directly for use by all federal departments and agencies. However, the committee believes that such an analysis can and should be developed by each federal department and agency and refined over time through repeated, consistent use by the relevant stakeholders and decision makers. At a minimum, a federally adapted business case analysis should explicitly include and clearly state the following: (1) the organization's mission; (2) the basis for the facility requirement; (3) the objectives to be met by the facility investment and its potential effect on the entire facilities portfolio; (4) performance measures for each objective to indicate how well objectives have been met; (5) identification and analysis of a full range of alternatives to meet the objectives, including the alternative of no action; (6) descriptions of the data, information, and judgments necessary to measure the anticipated performance of the alternatives; (7) a list of the value judgments (i.e., value trade-offs) made to balance achievement on competing objectives; (8) a rationale for the overall evaluation of the alternatives using the information above;

(9) strategies for exiting the investment; and (10) the names of the individuals and operating units responsible for the analysis and accountable for the proposed facility's subsequent performance. The business case format to be used by the department or agency should be agreed to by the pertinent oversight constituencies in Congress, the Office of Management and Budget, and the GAO.

RECOMMENDATION 4(b). To promote more effective communication and understanding, each federal department and agency should develop a common terminology agreed upon with its oversight constituencies for use in facilities investment deliberations. In addition, each should train its asset management staff to effectively communicate with groups such as congressional committees having widely different sets of objectives and values. Mirroring this, oversight constituencies should have the capacity and skills to understand the physical aspects of facilities management as practiced in the field.

Engineers, lawyers, accountants, economists, technologists, military personnel, senior executives, and elected officials lack a common vocabulary and style of interaction and do not necessarily share a common set of interests or time frames they consider important. To improve communications among the various stakeholders in facilities investments, each federal department or agency, in collaboration with the appropriate program examiners and congressional representatives, should develop and consistently use a common terminology for the concepts routinely used in facilities investment decision making and applicable to its organizational culture. With the wide variety of missions, cultures, and procedures that exist among federal departments and agencies, a standard set of government-wide definitions is not to be expected.

Training is necessary to ensure that the concepts underlying the terms have meaning and are understood by all. Facilities asset management staff should have the capacity and skills to understand the relationship of facilities to the big picture of an organization's overall mission and to communicate that understanding to others. They should also be able to solve problems by considering all sides of issues and to negotiate a solution that will best meet the organizational requirement. Financial, budget, and program analysts should receive some basic training in facilities investment and management.

RECOMMENDATION 5(a). Each federal department and agency should use life-cycle costing for all significant facilities investment decisions to better inform decision makers about the full costs of a proposed investment. A life-cycle cost analysis should be completed for (1) a full range of facilities investment alternatives, (2) the staff, equipment, and technologies inherent to the alternatives, and (3) the costs of the required funding.

For some very expensive project proposals, federal departments and agencies conduct life-cycle analyses internally to understand the total costs and benefits of the facility itself over the long term and to prioritize their requests for funding. However, in its research and interviews, the committee was not made aware of any instance in which a department or agency also conducted a life-cycle analysis for the staffing, equipment, and technologies inherent to the proposal, or for the life-cycle costs of the required funding.

RECOMMENDATION 5(b). Congress and the administration should jointly lead an effort to revise the budget scorekeeping rules to support facilities investments that are cost-effective in the long term and recognize a full range of costs and benefits, both quantitative and qualitative.

Under federal budget scorekeeping procedures, the budget authority associated with requests to design and construct a new facility, to fund the major renovation of an existing facility, to purchase a facility outright, or to fund operating and capital leases is "scored" up front in the year requested, even though the actual costs may be incurred over several years.

Scoring facilities' costs up front is intended to provide the transparency needed for effective congressional and public oversight. However, implementation of the budget scorekeeping procedures as they relate to facilities investments has resulted in some unintended consequences, including disincentives for cost-effective, long-term decision making and some gamesmanship.

Amending the scorekeeping rules such that they meet congressional oversight objectives for transparency and take into account the long-term interests of departments, agencies, and the public will not be easy. Amending them specifically to account only for life-cycle costs would probably create an even greater disincentive for facilities investments. The committee believes that a collaborative effort that encompasses a wide range of objectives, goals, and values is required. Some possible revisions to the rules could be tested through pilot projects.

RECOMMENDATION 6. Every major facility proposal should include the strategy and costs for exiting the investment as part of its business case analysis. The development and evaluation of exit strategies during the programming process will provide insight into the potential long-term consequences for the organization, help to identify ways to mitigate the consequences, and help to reduce life-cycle costs.

The development of exit strategies for facilities investment alternatives as part of a business case analysis will help federal decision makers to better understand the potential consequences of the alternative approaches. Evaluation of exit strategies can provide a basis for determining whether it is best to own or lease the required space in a particular situation and whether specialized or more generic "flexible" space is the best solution to meet the requirement. For those investment proposals in which the only exit strategy is demolition and cleanup, evaluating the costs of disposal may lead to better decisions about the design of the facility, its location, and the choice of materials, resulting in lower life-cycle costs.

RECOMMENDATION 7. Each federal department and agency should base its decisions to own or lease facilities on the level of control desired and on the planning horizon for the function, which may not be the same as the life of the facility.

Based on the committee's interviews and research activities, the criteria that departments and agencies use to determine if it is more cost-effective to own or lease facilities to support a given function are not clear or uniform. The committee believes that federal departments and agencies should base the "own" versus "lease" decision on a clearly stated rationale linked to support of the organizational mission, the level of control desired, and the planning horizon for the function to be supported.

RECOMMENDATION 8. Each federal department and agency should use performance measures in conjunction with both periodic and continuous long-term feedback and evaluation of investment decisions to monitor and control investments, measure the outcomes of facilities investment decisions, improve decision-making processes, and enhance organizational accountability.

Because the results of many federal programs or services are qualitative and occur over long periods of time, measuring them can be challenging. However, efforts are under way in various departments and agencies to develop indices and measures that can be applied to evaluate various aspects of facilities portfolios. Some or all of these indices could be adapted for use by other federal departments and agencies and used in combination with other metrics to measure the performance of their facilities' portfolios.

Short-term feedback procedures for facility projects are commonly used. However, to the committee's knowledge, no federal department or agency collected long-term feedback to determine if facilities investments met overall organizational objectives, solved operational problems, or reduced long-term operating costs. Long-term feedback is essential if the outcomes of facilities investments and management processes are to be measured and the decision-making process itself is to be improved.

RECOMMENDATION 9. To increase the transparency of its decision-making process and to enhance accountability, each federal department and agency should develop a decision process diagram that illustrates the many interfaces and points at which decisions about facilities investments are made and the parties responsible for those decisions. Implementation of facilities asset management approaches and consistent use

of business case analyses will further enhance organizational accountability.

In the federal government, responsibility and authority for making decisions and executing programs often are not directly linked. Instead, decision-making authority and decision-making responsibility are spread throughout the executive and legislative branches, leading to lack of clear-cut accountability for facilities investment outcomes.

A diagram that illustrates the many interfaces and decision points among the various federal decision-making and operating groups involved in facilities investment decision making can serve as a first step toward increasing the transparency of the process and enhancing accountability. Implementation of a facilities asset management approach, the use of performance measures and feedback processes, and the consistent use of business case analyses will further enhance organizational accountability for federal facilities investments.

RECOMMENDATION 10. Congress and the administration and federal departments and agencies should institute appropriate incentives to reward operating units and individuals who develop and use innovative and cost-effective strategies, procedures, or programs for facilities asset management.

In the federal system, the multiple-objective nature of laws and policies and the sheer volume of procedures sometimes result in unintended consequences, sometimes creating disincentives for good decision making and cost-effective behavior. Potential incentives to support more cost-effective decision making and management by facilities asset management groups could include programs that allow savings from one area of operations to be applied to needs in another area, if the savings are carefully documented; allow the carryover of unobligated funds from one fiscal year to the next for capital improvements, if doing so can be shown to be cost-effective; and establish meaningful awards for operating units with high levels of performance.

RECOMMENDATION 11 (from Chapter 5). In order to leverage funding, Congress and the administration should encourage and allow more widespread use of alternative approaches for acquiring facilities, such as public-private partnerships and capital acquisition funds.

A number of alternative approaches for acquiring facilities are being used by federal departments and agencies, on a case-by-case basis under agency-specific legislation. Each approach has advantages and disadvantages for particular types of organizations and types of facilities. None of the identified alternative approaches can guarantee effective management absent agreed-upon performance measures, feedback procedures, and well-trained staff.

Allowing the use of alternative approaches on a government-wide basis raises concerns about the transparency of funding relationships and concerns about whether the approaches sufficiently account for the perspectives of state and local governments and constituencies. Despite these concerns the committee supports more widespread use of alternative approaches to leverage funding and supports using pilot programs to test the effectiveness of various approaches and to evaluate their outcomes from national, state, and local perspectives. If changes to the budget scorekeeping rules are required to expand the range of alternative approaches, such changes should be tested through the pilot programs.

AN OVERALL STRATEGY FOR IMPLEMENTATION

Transforming decision-making processes, outcomes, and the decision-making environment for federal facilities investments will require sponsorship, leadership, and a commitment of time and resources from many people at all levels of government and from some people outside the government. Implementation of some of the committee's recommendations can begin immediately within federal departments and agencies that invest in and manage significant portfolios of facilities. However, implementing an overall framework of principles and policies will require collaborative, continuing, and concerted efforts among the various legislative and executive branch decision makers and operating groups. These include the President and Congress, senior departmental and agency executives, facilities program managers, operations staff, and budget and management analysts within departments and agencies and from the Congressional Budget Office, the Office of Management and Budget, and the GAO.

Having noted this, the committee is well aware that similar recommendations made by other learned panels advocating long-term, life-cycle stewardship of facilities and infrastructure have achieved only limited success and have failed to move all of the involved stakeholders to action. *The committee believes that a new dynamic can and must be instituted and recommends herewith a program it believes practicable.*

RECOMMENDED IMPLEMENTATION STRATEGY: The committee recommends that legislation be enacted and executive orders be issued that would do two things:

(1) Establish an executive-level commission with representatives from the private sector, academia, and the federal government to determine how the identified principles and policies can be applied in the federal government to improve the outcomes of decision-making and management processes for federal facilities investments within a time certain. The executive-level commission should include representatives from nonfederal organizations acknowledged as leaders in managing large organizations, finance, engineering, facilities asset management, and other appropriate areas. The commission should also include representatives of Congress, federal agencies with large portfolios of facilities, oversight agencies, and others as appropriate. The commission should be tasked to gather relevant information from inside and outside the federal government; hold public hearings; and submit a report to the President and Congress outlining its recommendations for change, an implementation plan, a timetable, and a feedback process for measuring, monitoring, and reporting on the results; *all within a time certain.*

(2) Concurrently establish department and agency working groups to collaborate with and provide recommendations to the executive-level commission for use in its deliberations. The working groups within each department and agency should collaborate with the executive-level commission. Staff in the departments and agencies are in the best position to communicate their organizational culture and identify practices for implementing the principles and policies that will work for *their* organization. In addition, they can provide the commission with information related to the characteristics of their facilities portfolios; issues related to aligning their portfolios with their missions; facilities investment trends; good or best practices for facilities investment and management; performance measures for monitoring and measuring the results of investments; and other relevant information.

The committee believes that such sponsorship, leadership, and commitment to this effort will result in

- Improved alignment between federal facilities portfolios and missions, to better support our nation's goals.
- Responsible stewardship of federal facilities and federal funds.
- Substantial savings in facilities investments and life-cycle costs.
- Better use of available resources—people, facilities, and funding.
- Creation of a collaborative environment for federal facilities investment decision making.

1

Context

BACKGROUND

The built environment in the United States is the result of several centuries of investment decisions about buildings and infrastructure. Generations of individuals and multitudes of public and private organizations have contributed to this evolving environment by making investments in the buildings (houses, offices, warehouses, factories, stores, museums, public safety stations, recreation centers, libraries, schools, hospitals, and research facilities) and infrastructure systems (water, waste disposal, energy, transportation, and telecommunications) that are the physical basis of our communities. This built environment and the services it provides directly affect the quality of life for more than 280 million U.S. residents as well as the strength of the national economy.

The magnitude of this investment is large. In 2000 the value of structures and utilities in the United States amounted to almost $22 trillion (USDOC, 2002). Seventy-seven per cent of these assets are owned by individuals, private, and not-for-profit organizations, while government (federal, state, local, and regional) owns about 23 percent (USDOC, 2002). And the investment is ongoing: Every year new facilities are built and existing ones are operated, maintained, and renovated.

The federal government also provides loans and grants to all 50 states and the District of Columbia, 38,000 local governments, and 36,000 special districts (U.S. Government, 2002) to finance the construction and operation of roads, transit systems, airports, housing, hospitals, schools, and utilities.[1] In 2001 such grants

---

[1] In addition to federal loans and grants, state and local governments raise funds through income, personal, and real property taxes and borrow money through bond sales repaid by these taxes.

and loans totaled approximately $145 billion (OMB, 2002). This report focuses on one aspect of the national investment in the built environment—the facilities that the federal government owns, leases, and operates directly.

To provide a context for Chapters 2 through 6, Chapter 1 describes the ongoing magnitude of the federal government's investment in facilities; reviews some fundamental characteristics of private-sector organizations and the federal government that affect facilities investment and management; and discusses drivers of change and conceptual shifts in facilities investment and management.

THE ONGOING INVESTMENT IN FEDERAL FACILITIES

As of September 2000, the federal government owned and leased approximately 3.3 billion square feet of space worldwide (GAO, 2003f). This space is distributed over more than 500,000 facilities, including military installations, courthouses, embassies, hospitals, administrative offices, museums, recreation complexes, and research campuses. The total value of federal facilities is conservatively estimated at $328 billion, with defense-related facilities accounting for about two-thirds of that total (GAO, 2003f). Annually, the federal government spends upwards of $21 billion for the direct acquisition of new facilities and the renovation of existing ones.[2] In fiscal year (FY) 2001, the federal government paid approximately $4.5 billion to power, heat, and cool its buildings (FEMP, 2003a). Federal agencies collectively spend more than $500 million per year for water and waste disposal (WBDG, 2003). Total government-wide expenditures for the operation, maintenance, repair, and disposal of federal facilities cannot be readily identified under the existing budget structure. However, annual expenditures are probably in the billions.

Figure 1.1 shows federal agencies' facilities holdings in millions of square feet as of September 2000. These figures do not include the 630 million acres of federal land holdings, including national parks, forests, and other uses, which make up 27.7 percent of the total land in the United States (USDOC, 2002).

Figure 1.2 shows the distribution of all types of facility space by use; infrastructure such as runways is not included. Office space, housing, and service space accounted for 60 percent of total federal government space (GAO, 2002b). The General Services Administration (GSA) owned or leased approximately 300 million square feet of the more than 728 million square feet of office space included in the federal inventory (GAO, 2002b).

Individual departments and agencies own and lease a wide range of facility types to shelter and support the people and equipment required to carry out their

---

[2]This figure is based on historic estimates. Line items for construction in the departmental appropriations bills were totaled for FY 2001.

CONTEXT

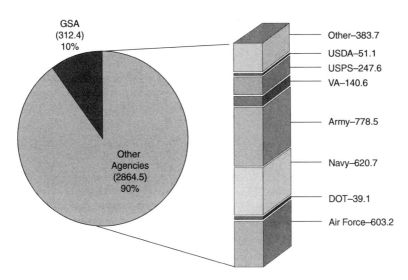

FIGURE 1.1 Federal agencies' facilities holdings in millions of square feet. SOURCE: GAO, 2001d.

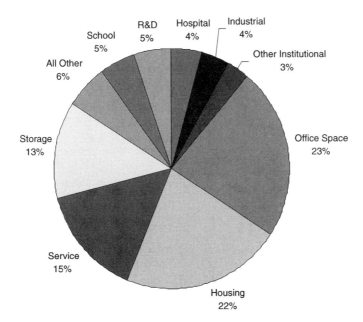

FIGURE 1.2 Distribution of federal government space by type of use. SOURCE: GAO, 2001d.

activities, programs, and missions. Some with narrowly focused missions—for example, the International Broadcasting Bureau and the Immigration and Naturalization Service—primarily use office space and a limited range of facility types such as radio transmission towers or border stations. The majority use specialized space—courthouses, embassies, museums, hospitals, prisons—in combination with office, warehousing, and research/laboratory space. The military services have the most diverse portfolios: Military installations contain all the types of facilities and infrastructure typically found in a small city, including airports, in addition to specialized facilities that support the defense mission.

SOME CHARACTERISTICS OF PRIVATE-SECTOR ORGANIZATIONS THAT AFFECT FACILITIES INVESTMENT AND MANAGEMENT

In the U.S. market economy, private-sector organizations are relied on to supply a wide variety of goods and services, and their activities are subject to regulation by many different governmental entities. Although the "private sector" is often referred to as if it is a monolithic entity, in actuality it is made up of tens of thousands of organizations with a myriad of purposes, operating with varying degrees of success. Some characteristics of private-sector organizations that affect their approaches to facilities investment and management are discussed below.

Mission and Goals

A private-sector organization is established to carry out a specific mission—its overriding "business." It is afforded latitude to achieve its mission through self-determined principles, policies, and practices within a public regulatory structure. Organizational missions are as wide ranging as the goods and services produced, from providing hospitality (hotel chains) and personal mobility (auto manufacturers), to solving complex business and technical issues for clients (consulting firms).

The goal of a private-sector organization, as opposed to its mission, is typically to achieve financial returns by selling goods and services at a higher price than the cost of producing them. A study of 146 multinational corporations found that 54 percent of the respondents chose "maximizing stockholder wealth" as their primary goal. The remaining respondents identified other goals, such as maximizing return on assets, maximizing growth in revenue, and maximizing growth in earnings per share (Block, 2000).[3]

---

[3]Other studies confirm this finding: Drury and Tayles (1997); Pike (1988); and the original, "classic" article by Mao (1970).

CONTEXT

For private-sector organizations, decisions to lease, own, build, renovate, renew, or dispose of facilities are driven primarily, but not exclusively, by market and financial considerations. Investments in facilities are made to ensure that operations are ongoing and efficient, a condition essential to the survival and growth of the organization in the marketplace. An organization's entire inventory of facilities typically is viewed and systematically managed as a "portfolio" of physical assets. Investments are made in these assets to support the organization's operational requirements.

Funding Facilities Investments

In 2001, U.S. businesses invested approximately $362 billion in new and existing structures and $748 billion in new equipment (U.S. Census Bureau, 2003). As illustrated in Figure 1.3, facilities typically account for almost one-quarter of a corporation's assets and its second or third highest operating cost (Brandt, 1994; O'Mara, 1999; Erdener, 2003), after people—salaries and benefits—and sometimes after technologies. New facilities or renovations of existing ones can cost tens to hundreds of millions of dollars, take two or more years to complete, and require annual investments for operations and maintenance over

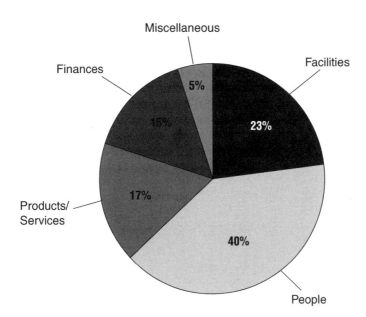

FIGURE 1.3 Distribution of total assets for a typical corporate organization. SOURCE: Adapted from Brandt, 1994.

a period of 30 years or longer. Millions of dollars may be spent annually to lease space.

Private-sector firms raise money for expenditures by (1) selling goods and services, (2) borrowing from a bank or other lender at a certain interest rate, and (3) selling stock in the company. When making investment decisions, they must look at the relationship between risk—the time uncertainty and volatility of a project—and returns—the expected receipts or cash flow (Groppelli and Nikbakht, 2000). The longer the cash flow is at risk, the greater the return must be. The value of financial capital must also be accounted for, because it changes over time: Money today is worth more than money in the future. Factors that influence the time value of money are inflation, risk (uncertainty of the future), and liquidity (how easily assets can be converted to cash).

Private-sector firms typically budget for two types of expenditures: operating and capital. Operating expenditures (e.g., wages, salaries, administrative, and other current costs) are short-term and are written off in the same year as they occur. Capital expenditures (e.g., buildings, equipment, patent rights) are long-term and are amortized over a period of years, as determined by tax regulations (Groppelli and Nikbakht, 2000). Budgets for both types of expenditures are linked by an overall management plan.

Private-sector organizations make decisions about capital expenditures separately from decisions about operating expenditures. Capital spending decisions are made based primarily on how they affect shareholders and are evaluated predominately in monetary terms (PCSCB, 1999). In capital decision making and budgeting, there is no such thing as a risk-free project, because future cash flows may decline at any time owing to inflation, loss of market share, increased costs for raw materials, labor, or other resources, new environmental regulations, or higher interest rates, among other factors.

When considering a potential facilities investment, private-sector standard practice is to first conduct a financial analysis. The analysis, embodied in a pro forma statement, typically evaluates the net present value (NPV) of the potential investment by projecting the revenues the investment is likely to generate, discounting the future cash flow by the time value of money, accounting for risk, and subtracting the initial costs. Under such a process, it makes economic sense to proceed with a more detailed evaluation of a facilities investment only if the NPV is positive. Facilities investment analyses, decision making, and evaluation processes are discussed in detail in Chapter 3.

Response to Change

In a competitive marketplace, the organizations that survive are those that can adapt to continual and often rapid change. For-profit organizations with long-term success are constantly modifying factors such as cost, availability, and the characteristics and qualities of goods and services to meet market conditions and to prepare themselves for meeting new competitors. They also tailor their multiple offerings of goods and services to fit specific market segments so as to realize the maximum yield (profits, short-term market share, or market segment control) for the dollars invested.[4]

As long as profits ensue, a private-sector organization's mission, values, and leadership can remain relatively unchanged for years. However, its principles, policies, and practices for meeting its mission may be adjusted continually or adapted in response to dynamic changes in the operating environment. Adjustments such as internal reorganization to eliminate unproductive overhead costs or to address underperforming business units may be necessary as a start-up business becomes a more mature, stable organization and as the scale of its operations grows or declines. When change requires the acquisition of new skills or access to newly developing markets, the acquisition of one company by another or the merger of two is not uncommon. For private-sector organizations, the issue frequently is not whether change is needed but when and how to change. Few elements are fixed in the drive to improve organizational performance in order to meet financial goals and achieve strategic objectives. Timing is critical since organizations that are slow to sense the need for change or to make adjustments are disadvantaged in the subsequent time period.

Flexibility

Successful private-sector organizations are able to respond to market or other changes relatively rapidly because they build flexibility into their decision-making processes, their procedures, their culture, and the strategies used for delivering and acquiring space. They use a mix of ownership, leasing, lease-purchase, and other financial arrangements to acquire facilities depending on how the space will be used to support their operational requirements.

Some private-sector organizations also build flexibility directly into their facilities: buildings with components and furniture that can be relatively easily reconfigured to accommodate new uses or new technologies, thereby allowing changes to be made in the physical environment relatively rapidly and at a relatively low cost. This is important in an environment where the turnover of employees can necessitate the reconfiguration of workspace on a 12- or 18-month (or shorter) cycle. Flexible facilities are also built as a hedge against change: If a facility is being built to meet a particular requirement and that requirement changes soon after the facility is operational, it can be adapted to other uses. Flexibility in design can also make a facility more marketable to other users if and when the organization chooses to sell it.

---

[4]For example, the Marriott Corporation has developed distinct lines of hotel accommodations differentiated by ownership, quality, level of service, and cost per night that can be matched to local markets.

As important, if not more important, to meeting the organizational mission are the people within the organization, the quality of the leadership and management, and the skills of the workforce. Private-sector organizations have considerable flexibility to adjust their workforce to achieve their organizational mission and goals. They can adjust their compensation packages to the market, offering high salaries and a range of benefits to attract those who possess the leadership, management, and technical skills required to execute the organization's core business lines. Within labor practice constraints, they can lay off workers in response to changing markets, mergers, or other factors and can dismiss on short notice those who do not perform satisfactorily.

SOME CHARACTERISTICS OF THE FEDERAL GOVERNMENT THAT AFFECT FACILITIES INVESTMENT AND MANAGEMENT

In addition to the President, Congress, and the judicial branch, the federal government's executive and legislative branches today comprise 15 departments, 40 independent agencies, 22 corporations and commissions, and approximately 1.7 million civilian employees (U.S. Government, 2003). This structure incorporates a system of checks and balances that ensures that many aspects and possible outcomes and consequences of policies and decisions are identified, considered, and accommodated in some fashion.

Decision-making authority and responsibility for establishing missions, objectives, policies, and practices are spread throughout the executive and legislative branches—the President and the Cabinet, the Congress, senior executives and a multitude of managers in operating and oversight agencies and, ultimately, the voting public. The judicial branch acts as another check on the system by ruling on the constitutionality of decisions made or actions taken.

Because of its size and organizational structure, the federal government does not act as a single, independent, monolithic entity. Instead, it operates more like a network of distinct but interdependent organizations with multiple missions, cultures, structures, and decision-making processes.

One distinction between nongovernmental and governmental organizations is the beneficiary of their respective investments in facilities and infrastructure. Nongovernmental organizations directly reap most of the benefits, or losses, from spending on their facilities, buildings, and equipment. When the federal government invests in facilities, the public at large benefits or loses. Investments that confer benefits on a wide class of parties are referred to by economists as "public goods," because no private person or firm can capture all of the benefits. Public goods and services are distributed universally, that is, to all segments of society regardless of whether it is economically efficient to do so.[5] Thus, decisions about federal facilities investments must take into account the benefits to the public at large, not just the benefits to a specific organization, agency, or department. In many cases, these benefits are nonfinancial in nature—for example, preservation of a historic structure.

However, it is difficult for the public at large to directly influence facilities investment decisions at the federal level.[6] Those who most directly influence federal facilities investment include department and agency senior executives, facilities program managers, budgeting and financial analysts, Congress, the President, other policy makers, and special interest constituencies. The President and Congress are responsible for providing leadership and vision, setting policies, enacting legislation, establishing regulations, and authorizing and appropriating public funds. Civil service employees and political appointees within the various federal departments and agencies are responsible for administering programs, establishing and executing processes, analyzing their results, recommending initiatives, enforcing regulations, and expending public funds efficiently, effectively, and legally.

In this decision-making structure, the various government entities have diverse but overlapping objectives. As shown in Figure 1.4, some decision-making and operating groups, such as the Office of Management and Budget (OMB) and the Congressional Budget Office (CBO), focus on government-wide issues, like balancing the budget. Departments and agencies focus on issues related to their specific missions.

Goals and Missions

At the highest level, the goal of the federal government is to promote the public's health, safety, and welfare. Individual agencies have specific missions designed to support achievement of this goal. Their missions include, but are not limited to, providing national defense and homeland security; conducting foreign policy; protecting wilderness areas, national parks, and national landmarks; conserving the nation's historical documents and cultural artifacts; supporting public education; and regulating businesses, transportation safety, and the quality of food, water, workplaces, and the environment. These missions are viewed as inherently governmental, although some of the activities of all of them are performed by private-sector organizations.[7]

---

[5]An example is the provision of mail delivery by the U.S. Postal Service to all residents, even in sparsely settled and isolated areas, where the per capita costs of providing such services result in operating losses. In addition, the price to the consumer of a first-class stamp is the same anywhere in the United States although the cost to the Postal Service for delivering a letter varies greatly, depending on distance and location.

[6]At state and local levels, the public can have a direct say in facility investment decisions by voting for or against bond referendums and by directly influencing the setting of tax rates.

[7]For example, although the government is responsible for providing national defense, it contracts with private-sector organizations to produce the weapons systems required to achieve that mission.

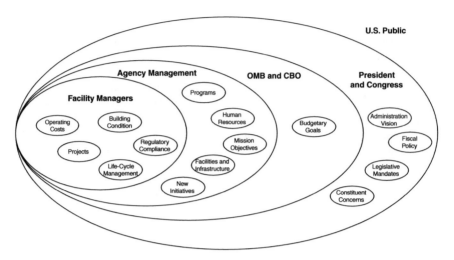

FIGURE 1.4 The various stakeholders in facilities investments and their diverse and overlapping objectives.

Funding Facilities Investments

The U.S. government primarily collects taxes and sells debt instruments, such as Treasury bonds and notes, to raise funds to support its activities.[8] All expenditures, both operating and capital, are accounted for in the annual Budget of the United States Government.[9] Since 1945, a number of actions and studies have been initiated to determine if the federal government should institute procedures to allow for separate consideration of operating and capital expenditures.[10] To date, such procedures have not been implemented, and the government continues to make decisions about and budget for operating and capital expenditures together, unlike private-sector organizations. Thus, facilities investment decision making in the federal government is driven in large part by the annual budget process and its associated time frames and procedures. These processes and procedures drive a short-term perspective, one that focuses on current expenditures as opposed to long-term investments.

Federal budgeting is a continual process that has specified milestone dates,

---

[8] The government also raises funds by charging for some services and leasing properties to outside interests.

[9] The federal government first instituted a central budget under the Budget and Accounting Act of 1921. Prior to 1921, federal departments and agencies submitted individual budgets to Congress.

[10] These initiatives include the Hoover Commission (1949); the President's Commission on Budget Concepts (1967); and the President's Commission to Study Capital Budgeting (1999).

usually annual, by which time a formal budget must be presented. Two parallel processes and two time cycles are at work: the Presidential budgeting process and the Congressional budgeting process and an annual cycle that meshes with tax reporting and appropriation cycles (the operating budget) and a longer-range (out-year) budget cycle that gives a better picture of where a department or agency is going beyond the current snapshot in time. Unlike private-sector organizations, which have some flexibility to internally establish their own budgeting and funding processes, all federal departments and agencies must comply with one government-wide set of budgeting procedures.

In the federal budget process, as in many private-sector enterprises, requests for funding typically exceed expected resources. Only a relatively small proportion of the federal annual operating budget is discretionary, because the bulk of it is constrained by prior agreements, such as entitlements, and by the need to maintain ongoing programs and services seen as critical or valuable. In any environment where expectations exceed resources, trade-offs must be made. Decision makers in Congress and federal departments and agencies are asked to balance the competing demands of very different programs: Funding for facilities investments must be weighed against funding for medical research, weapons systems, homeland security, education, or numerous other public programs.[11]

In many cases, therefore, federal policy and budget decisions are fundamentally matters of achieving political consensus. Where a difficult decision is at stake, the government often operates on the principle that, absent a clear consensus, it is better not to act but rather to continue to seek a consensus. The governmental process is not one that chooses to settle on one or another proposal based solely on a financial or technical analysis. Instead it seeks to fashion a compromise proposal that will command the greatest degree of consensus from among those offered. In this operating environment, programs or investments whose results are not highly visible or will only be realized in the long term, such as facilities maintenance, tend to be put off to out-year budgets.

Response to Change

The federal government is less driven to change or to adapt its operating principles, policies, or practices or its organizational structure than is the private sector. Change or adaptation in government is not driven by market forces but is more likely to occur in response to elections, major events, socioeconomic trends at home and abroad, budget projections, media attention, outside or internal evaluations of agency performance, or changes in perceived good management practices.

---

[11]Private-sector organizations are rarely involved in making trade-offs among such disparate demands.

Typically, change occurs slowly, except perhaps in isolated cases during crisis situations. The system of checks and balances guards against constant or rapid change and upheaval in the delivery of public goods and services. Consensus building to make a change can take years and span several election cycles because of the many vested interests involved—elected leaders, Congressional committees, agency staff, contractors who work for the government, and the public. The reorganization of departments and agencies or the divesting of government programs is typically a lengthy and controversial process. However it is possible and can be done when the need is clear.[12] For these reasons and others, the missions of the federal government, its departments, and agencies often remain relatively unchanged at strategic levels for long periods of time, although many management practices change over time and the missions of individual agencies do evolve.[13]

Flexibility

The scale of government operations is invariably large and typically precludes flexibility and scalability. The government is frequently a monopoly provider of goods and services, either because the function is inherently governmental or because of legislation. In some instances, the government's role is to develop products and services initially and then spin them off to the private sector once the feasibility has been established and risk factors have been understood.[14]

Because of the federal government's size and other factors, most of its activities are governed by numerous procedures that are designed to achieve some uniformity in the use of and accounting for resources. Such procedures limit the flexibility that can be applied to operations, including the hiring and firing of the workforce.

Leadership in the federal government is primarily provided by the President, the Congress, the Cabinet, and other high-level political appointees. In contrast to the private sector, the election process may cause constant change in leadership. Each administration establishes a vision of the future and puts forward strategies for achieving that vision. However, it does not have the flexibility to unilaterally implement those strategies but must either work within established procedures

---

[12]Examples include the separation of the regulatory and advocacy functions of the Atomic Energy Commission (now the Department of Energy), the establishment of a Department of Homeland Security, and the divestiture of some aspects of the communication satellite business.

[13]The Department of Energy (DOE), for example, was originally established as the Atomic Energy Commission in the 1940s to produce nuclear weapons. Today, in addition to the nuclear stockpile, DOE's missions focus on energy production and conservation and the cleanup of waste from the weapons programs.

[14]Space-based systems for communication and Earth imaging are cases in point.

and processes or enact legislation to change those procedures, typically a time-consuming and difficult process.

Civil service employees, whose tenure is not dependent on the political party in office, carry out federal government programs and initiatives. Federal departments and agencies must seek to attract workers with the required management and technical skills by using relatively standardized compensation packages with clearly established salary ranges and salary caps. Their ability to adjust their workforces to meet changing requirements is similarly limited in that it is a time-consuming process to reassign or lay off workers whose skills are no longer essential to the achievement of the mission. Dismissal of civil service employees for unsatisfactory performance can also be a difficult and lengthy process. As a catalyst for workforce restructuring, federal agencies have repeatedly been given "buyout" authority in recent years. Such authority provides financial incentives for individuals to retire or seek work elsewhere, allowing some adjustments in the size of the workforce and the allocation of positions.

The lack of flexibility in processes and procedures also applies to most facilities. Historically, federal departments and agencies acquired the majority of their facilities on a one-off basis—that is, a facility was designed to serve a specific purpose or function; such facilities include courthouses, embassies, research laboratories, museums, and hospitals. In addition, many federal buildings are historic in character and require specialized renovation techniques. Because most federal facilities are used for 50 years or longer, many of them must be adapted to support new functions when requirements change: A former barracks might be reconfigured for use as administrative space.

Efforts such as GSA's integrated workplace are intended to provide more flexibility in building systems and components so that they can be more easily adapted to changing requirements and technology (GSA, 1999). However, the vast majority of federal facilities were clearly not designed for flexibility and are difficult and expensive to reconfigure or adaptively reuse in response to changing requirements.

FACILITIES REQUIREMENTS, LONGEVITY, AND LIFE-CYCLE COSTS

The last two decades have brought great changes in the way Americans live and the services they demand. External and internal forces such as radical advances in computers and communication, the regulatory environment, changes in demographics and socioeconomic conditions, and a renewed emphasis on the safety of personnel and customers are driving change in the operating environments of all types of organizations.

Today, organizations can operate around the clock by having business units located around the world and networked through technology. An increasingly diverse workforce requires greater accessibility and work arrangements such as telecommuting, flexible or part-time schedules, child care, and the like. Technologies such as the Internet and wireless connectivity are changing the ways in which the public accesses services and the ways in which organizations interact with their employees, customers, and clients. Because electronic communication allows for the rapid exchange of information and the rapid collection and tabulation of demands and viewpoints, it provides ways to increase the number of participants in the marketplace and in public processes. All of these changes have an effect on facilities requirements, design, and operations.

Facilities Requirements

Changes in society and in organizational environments affect facilities requirements—that is, the properties of a facility that will achieve a balance between the external environment, the facility's long- and short-term objectives, and the functions it is expected to serve (Erdener, 2003). Twenty-four-hour operations, together with computers and other office equipment, make the uninterrupted supply of a facility's power, heating, ventilation, and air-conditioning systems more critical. Increasing turnover rates for employees and operating units and new technologies for business and security functions necessitate facilities that can be adapted to new interior layouts quickly, efficiently, and cost-effectively. An increased emphasis on physical security calls for methods to reduce the vulnerability of a facility to terrorism and natural hazards so as to better protect the people and equipment inside. Greater accessibility for the physically handicapped, employees, and visitors requires new facilities designs.

Facilities' Longevity

Although facilities requirements are dynamic, facilities themselves are relatively static and can be long-lived. Most facilities are designed to provide at least 30 years of service. With proper maintenance and management, buildings can perform adequately for 100 years or longer.

The longevity of an individual facility is dependent on such factors as quality of design; quality of construction; durability of construction materials and component systems; incorporated technology; location and local climate; type and intensity of use; operation and maintenance methods; damage caused by natural and man-made disasters; and human error (NRC, 1998; FFC, 2001c). Facility longevity is also influenced by its value to its owner: A facility that is performing adequately may still be demolished if it no longer fulfills an organization's operating requirements, or if other opportunities for the land on which it is situated provide greater value.

Thus, a central issue in addressing facilities investments is the relative longevity of facilities and the likelihood that whatever is built and however it is maintained will eventually become obsolete to the original objectives in the short, intermediate, or long term. There will also be changes in ownership, occupancy, regulations, condition, internal and external technologies, and the opportunities that inhere in a facility or the real estate it occupies.

Life Cycles of Facilities

Facilities are complex structures with a number of separate but interrelated systems—exterior walls, roofs, and windows; mechanical and electrical systems; heating, ventilation, and air conditioning; fire protection; security; and others. The individual systems require extensive renewal periodically, on cycles that vary from 10 to 50 years. Because a facility's systems can be repaired or replaced and its interior spaces can be reconfigured to support new functions, its service life can be extended well beyond the life of the individual systems. For this reason, facilities can be viewed as renewable assets.

Facilities pass through a number of stages during their lifetimes: planning (programming, conceptual planning, design), acquisition (construction, start-up), operation (use, renewal, repair or revitalization), and disposal (sale, demolition) (Figure 1.5). The direct costs of facilities over their life cycle include those for programming; conceptual planning; financing; design; construction; maintenance; repairs; replacements; alterations; normal operations such as heating, cooling, lighting; and disposal.

Design and construction expenditures, the so-called "first costs" of a facility, typically account for 5 to10 percent of the total life-cycle costs. However, decisions made during design and construction about how much to invest in a building's materials and systems can significantly impact its operating and "exit" or disposal costs. Operation and maintenance costs typically are 60 to 85 percent of the total life-cycle costs, with land acquisition, programming, conceptual planning, renewal or revitalization, and disposal accounting for the remaining 5 to 35 percent (NRC, 1998; FFC, 2001c). For facilities to perform adequately and reach their design service life, annual investments in preventive maintenance and minor repairs are required.

Facilities, of course, are built to shelter people and equipment and their activities. Thus, in addition to the cost of the facility itself, there will be related costs such as those for staffing, furnishings, equipment, and information technologies. These costs can be 2 to 10 times greater than the cost of the facility over its entire life cycle.

FIGURE 1.5 Facility life cycle.

As noted previously, private-sector organizations invest in facilities to ensure that the production of goods and services and other operations are efficient and ongoing in order to maximize their returns. When public-sector organizations face choices on where to invest limited resources, facilities investments, particularly investments in maintenance and repairs, are often the first to be deferred or cut altogether. For public-sector officials, this decision is relatively easy, because in the short term operations will continue without an obvious immediate decline in services to the general public. As maintenance is deferred over the longer term, however, the capital investment required to renew or replace a facility is twofold: the replacement cost and the return on the original investment. It has been estimated that the cost relationship is between $4 and $5 in capital liability created for each $1 of deferred maintenance (Kadamus, 2003). Thus an accumulation of deferred investments over the long term may be significantly greater than the short-term savings that public-sector decision makers were initially seeking.

CONCEPTUAL SHIFTS IN FACILITIES INVESTMENT DECISION MAKING

In the drive to achieve their missions, increase profitability, and become more competitive, private-sector organizations have sought to significantly improve critical areas of performance by being results-driven. They focus on improving operations linked to financial, functional, and corporate objectives such as increased yields, reduced delivery times, increased inventory turns, improved customer satisfaction, [and] reduced product development time (Schaffer and Thomson, 1992).

Research reports published in the 1980s and early 1990s found that finance directors and corporate planners responsible for the business planning and direction of private-sector organizations were not closely linked to their facilities management or real estate departments (Then, 2003). As competition and the pressure to produce results increased in the 1990s, financial directors and corporate planners began scrutinizing all of the costs of doing business, including facilities costs, in order to remain competitive. Some organizations began to take a more integrated, all-encompassing approach to managing their resources—people, facilities, information technology, and dollars—to better meet their missions. When senior managers recognized that the facilities required to support the delivery of goods and services were a means to a more basic economic end, their organizations began to evaluate facilities investment proposals as they would proposals for other investments—as mission-enablers rather than solely as costs. In this construct, investments in facilities and decisions on their location are typically made to ensure that business operations are continuous and efficient, essential ingredients to an organization's current and future success.

The emergence of new information technologies in the 1990s also drove and enabled more integrated approaches to facilities investments and management, although such technologies also present organizational challenges. Using information technology for facilities management is not new: Computer-aided facilities management systems have been available for almost two decades. What is new is the capacity to integrate data from facilities management systems with data from financial and personnel systems in order to track all of the resources involved and provide the information needed to make decisions about investments. These technologies also allow for the rapid aggregation of large amounts of data from geographically dispersed sites. Thus, data can now be gathered for entire portfolios of facilities and their staffing and operating costs as opposed to data for individual buildings only. At the same time, determining which data are actually useful in decision making can be difficult. Doing so is likely to require a concerted effort to identify, verify, and refine data in order to develop information that is helpful in differentiating the consequences of alternative actions.

All of the above factors—the desire for flexibility, responsiveness to change, changing expectations, integrated management, information technologies—are driving significant change in the field of facilities management. The evolving discipline of facilities asset management is the focus of Chapter 2.

2

Facilities Asset Management

BACKGROUND

The field of facilities management is evolving. Once focused on tactical concerns, tasks, and functions that were oriented to the operation of individual buildings, it now focuses on the entire portfolio of facilities and integrated resource management (Figure 2.1).

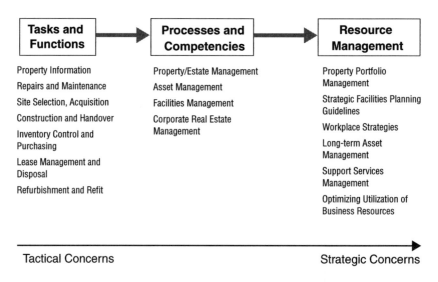

FIGURE 2.1 The evolving focus of facilities asset management. SOURCE: Then, 1996; 2003.

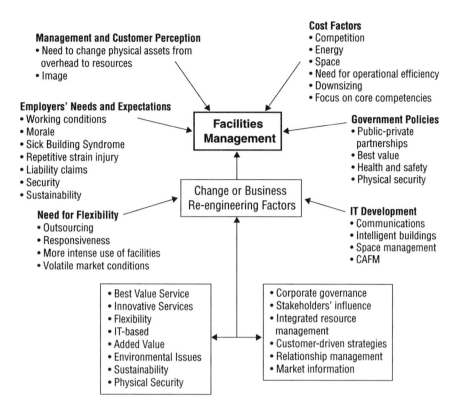

FIGURE 2.2 Factors driving the evolution of facilities management. SOURCE: Adapted from Okoroh et al., 2002; 2003.

One driver of this evolution is the emphasis by private-sector organizations on results-driven management strategies for all aspects of their operations. This shift is also indicative of the increasing recognition of facilities as "mission enablers" that support organizational goals, work processes, and productivity. Increased competition, a renewed emphasis on physical security, the outsourcing of business functions, changing expectations and requirements of employees and clients, and emerging information and building technologies are also factors (Figure 2.2). As noted by corporate real estate expert Martha O'Mara,

> the organizational emphasis of corporate real estate is shifting from a functional project management approach based on how buildings are delivered to one which aligns with the structure of the company and the way work is conducted. This shift is necessitated not only by the strategic perspective but also by the increased use of service providers outside of the company that assume many of the routine functions of real estate and facility management (O'Mara, 1999, p. 307).

While many factors are driving the evolution of facilities management, new technologies are enabling it. The development of open platforms and relational databases allows for the integration of data from disparate sources, including financial, facilities, and personnel systems. Large quantities of data from geographically dispersed locations can be gathered and processed quickly to monitor day-to-day operations, costs, and trends. Decision support tools allow for the development and evaluation of large numbers of alternative investment scenarios.

The next sections focus on the emerging practice of facilities asset management, its components, and the additional skills required of facilities asset managers. Chapter 2 concludes with a summary of principles and policies from best-practice organizations.

FACILITIES ASSET MANAGEMENT

Facilities asset management is an evolving discipline. In this report it is defined as "a systematic process of maintaining, upgrading, and operating physical assets cost-effectively. It combines engineering principles with sound business practices and economic theory, and provides tools to facilitate a more organized, logical approach to decision making" (FHWA, 1999, p. 7). A facilities asset management approach allows for both program- or network-level management and project-level management and thereby supports both executive-level and field-level decision making.

Program- or network-level management is associated with a systemwide approach that involves structured decision-making practices, including the analysis of trade-offs to identify and execute the best investments for a portfolio of facilities. Such management involves a macroscopic view of the assets being managed and makes use of aggregated data. Project-level management decisions, in contrast, are associated with identifying the best actions to take for specific facilities, and they typically occur at the field level, using more disaggregated data (Figure 2.3).

The importance of a facilities asset management approach is that it allows organizations to integrate facilities considerations into corporate decision making and strategic planning processes. This is a significant shift from past practice, whereby facilities-related decisions were often made after the organization's strategic direction had been set. Using a facilities asset management approach allows organizations to forge a direct link between organizational goals, facilities investment decisions, and day-to-day operations (Figure 2.4).

COMPONENTS OF A FACILITIES ASSET MANAGEMENT APPROACH

Facilities asset management is different from asset management in the financial/legal sense for the following reasons (Tracy, 2001):

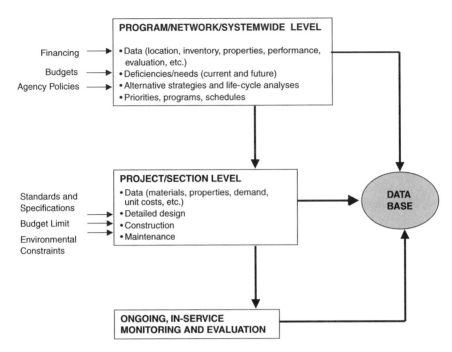

FIGURE 2.3 Components of a facilities asset management system. SOURCE: Adapted from Hudson et al., 1997.

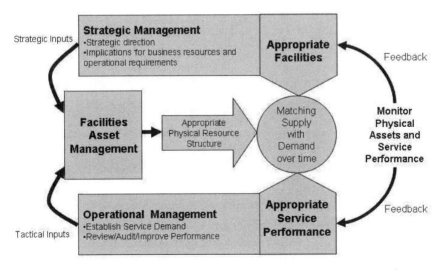

FIGURE 2.4 Linking organizational goals with facilities investment and operations. SOURCE: Adapted from Then, 1996; 2003.

- It focuses on a subset of corporate balance sheets that are physical in nature.
- It enlarges the scope of assets to extend to noncapitalized assets such as leased space, office equipment, human resources, and the like.
- It includes operating assets that require regular maintenance and repair to retain their functionality and avoid catastrophic failure.
- It includes operating assets that depreciate and wear out over time but that can also be renewed through investment (renewable assets).
- It involves the use and deployment of assets in dispersed locations and over the various operating units of an organization.
- It accounts for a return on investment that is often in the form of increased productivity of a facility's occupants, a difficult value to quantify and measure.
- It recognizes that the facilities program manager may or may not have authority over the disposition of all or a portion of the assets that he or she manages.

The literature on facilities asset management identifies several components needed to ensure that investment decisions are aligned with the mission and goals of an organization:

- Accurate data for the entire facilities portfolio, not just individual buildings, to enable life-cycle decision making.
- Models for predicting the future condition and performance obtainable from these facilities as a portfolio.
- Engineering and economic decision support tools for analyzing trade-offs among competing investment approaches.
- Performance measures to evaluate the impacts of different types of actions (e.g., maintenance versus rehabilitation) as well as the timing of investments on the overall goals for service provision.
- Continuous feedback procedures.

These components are described in greater detail below.

Accurate Data

Facilities asset management data at a minimum include inventory and attribute data. Inventory data describe elements of assets that do not change as a function of time—for example, the number, location, type, and size of facilities and the year of acquisition. Inventory data are gathered in a relatively straightforward manner, even for large portfolios of facilities; once gathered, the time and cost to update them are minimal.

Attribute data capture characteristics that do change over time, such as the demand for the facilities, usage, value, age, maintenance history (including treatment types and timing), operating and repair costs, condition, and so forth. Attribute data are more difficult to gather initially than are inventory data. Updating attribute data may require periodic condition assessment and other programs, which can be costly. Computer-aided facilities management systems are used to store, analyze, and update both inventory and attribute data.

Performance-Prediction Models

Performance-prediction models predict the deterioration of building components, measured as a composite condition index, as a function of time.[1] They are important because certain components of a facility are particularly prone to deterioration or failure and require relatively frequent maintenance or repairs. Some mechanical and electrical systems of a facility tend to have numerous moving parts and are likely to need a great deal of maintenance (they are said to have a high maintenance need probability). Nonperformance of some of these components can have serious consequences for the serviceability of the facility. Similarly, life safety systems generally have interacting parts, such as electrical signal systems or controls, that have both a high maintenance need probability and very serious consequences if they do not perform properly.

Building envelopes, for example, may have a relatively high maintenance need probability, and the effects of nonperformance can range from annoying to catastrophic. The envelope's exposure to the weather makes it vulnerable, and hidden deterioration may result if leaks are unknown or neglected. In contrast, the covered structural system of a building tends to remain unaffected for the life of the facility (a low maintenance need probability) unless the loading is significantly changed, or the structure is modified, or deterioration occurs.[2] The consequence of nonperformance of a structural element is almost always serious. Having models that can help to identify the differing maintenance need probabilities of facilities can help facility managers and others determine where resources can be spent to achieve the most significant returns in terms of supporting the organization's operations.

Engineering and Economic Decision Support Tools

Engineering-economic ranking and optimization methods can help decision makers to evaluate trade-offs among different investment approaches. Ranking

---

[1]The BUILDER system developed by the Construction Engineering Research Laboratories of the U.S. Army Corps of Engineers is one example of a performance prediction model; other models have been developed by private-sector software firms.

[2]One example of an uncovered structural system is a steel bridge, which has a high maintenance need probability.

methods make use of various decision criteria to prioritize competing needs in an overall facility portfolio or infrastructure system. Decision criteria might include the current condition, the predicted condition at some future time, life-cycle costs, cost-effectiveness (i.e., some measure of effectiveness per unit cost of improvement), and benefit-cost ratio. The relative sophistication of the decision criteria used for project rankings ultimately impacts the relative value gained per unit investment.

Optimization methods such as mathematical programming methods are used to identify the combination of competing investment options that would result in the greatest return on the investment, given budget constraints. Although several attributes of the facility or system investment may be quantifiable as benefits or costs, not all such attributes are quantifiable—for example, the environmental and social impacts of various facility investment decisions. Such attributes may, however, be considered qualitatively in various multiattribute decision frameworks.

Performance Measures

Performance-prediction models to project what may happen are an important element of a facilities asset management approach. Equally important are performance measures to gauge what has occurred or is occurring in respect to a facilities-related operation or activity.

Most organizations, whether private or public, measure the performance of individual projects or buildings. Typical indicators include project completion in relation to the original schedule and budget; energy, utility, or other operating costs per square foot; utilization rate (occupied space as a proportion of usable area); facility condition; and the like.

Indicators to measure the performance of an entire portfolio of facilities in relation to organizational goals are less well developed but are fundamental to a program- or network-level management approach. Performance measures in general and program-level indicators specifically are discussed in greater detail in Chapter 4.

Continuous Feedback

One of the objectives of implementing a facilities asset management approach is to ensure the alignment of an organization's portfolio of facilities with its mission and operating objectives. Continuous feedback is required to monitor the operating condition of facilities that directly support and impact organizational mission; to identify facilities that are no longer needed due to changing requirements; and to identify facilities that are obsolete technologically or otherwise. This information, in turn, can be used to determine where investments should be made to acquire, renew, or dispose of facilities. Continuous feedback and monitoring are discussed in greater detail in Chapter 4.

FACILITIES ASSET MANAGERS

The usefulness of a facilities asset management system is closely tied to the extent to which an asset management culture has permeated the organization, the quality of data on the asset portfolio, the linkage between the asset management goals and organizational mission, and the skill level of the people involved in the management system. Implementing a facilities asset management approach also requires that facilities staff at headquarters and in the field have the appropriate background and training to provide strategic information and to make recommendations to senior managers.

The importance of having a competent workforce with the appropriate skills and training to support an organization's core competencies, goals, and missions cannot be overestimated. A recent study of private-sector organizations that were able to "make the leap from good to great" and to sustain their results for at least 15 years found that:

> "Who" questions come before "what" decisions—before vision, before strategy, before organizational structure, before tactics. *First* who, *then* what—as a rigorous discipline, consistently applied . . . . The old adage "People are your most important asset" is wrong. People are not your most important asset. The *right* people are . . . . Whether someone is the "right person" has more to do with character traits and innate capabilities than with specific knowledge, background or skills. (Collins, 2001, p. 63)

Thus, people, like facilities, technologies, and dollars, are mission enablers, assets that must be invested in over time.

In a facilities asset management approach, facilities managers can no longer be regarded only as caretakers who bring unwelcome news about deteriorating facilities and the need for investments. As facilities management has evolved from tactical, building-oriented activities to a strategic, portfolio-based approach, the skills required by facilities management organizations have similarly evolved. A facilities asset management approach requires not only the technical skills (e.g., engineering, architecture, mechanical, electrical, contracting) found in traditional facilities engineering organizations but also business acumen and communication skills.

A report by the Center for Construction Industry Studies (CCIS) involving 31 private and public sector organizations found that it is fairly well recognized in owner firms that the skill set required to manage and work on projects from the owner's side has changed dramatically and the issue of skill development of owner personnel is perhaps the most difficult one facing owner firms (CCIS, 1999). In business terms, critical owner skills include technical knowledge of the process, alignment with the business units' goals and objectives, facility definition, stewardship of the overall project process and objectives, and project controls (Sloan Program for the Construction Industry, 1998). Skills required by facilities asset managers are outlined in Table 2.1.

A newly released study reinforces the CCIS report and lists 27 business skills a facility manager should have to be effective in today's operating environment (Table 2.2). According to the International Facility Management Association (IFMA), only 34 percent of facility managers have business degrees (IFMA, 1998). Thus,

> [it] is not surprising that facility managers are unsophisticated in applying business practices to facility management. Most of them have technical education in engineering, architecture, or administrative management. Their education and training did not stress business principles or theory. Many of them have little training in financial management. (Cotts and Rondeau, 2004, p.3)

For these reasons, most organizations adopting a facilities asset management approach must have staff who are able to use new methods of analysis, who understand financial concepts and management, and who can communicate ef-

TABLE 2.1 Skills Required by Facilities Asset Managers

Category	Skill
Business	Writing and managing contracts Negotiation Managing budgets and schedules
Communication	Coordination/liaison Conflict management Cultivate broad network of relationships
Influence	Mentoring Motivating Change management
Managerial	Team building Delegating Politically aware/see big picture
Problem solving	Continually analyze options/innovation Planning Consider all sides of issues, risk management
Technical	Understand entire construction process Multidisciplined (knowledge of several areas of engineering) Information technology skills

SOURCE: CCIS, 1999.

TABLE 2.2 Business Skills for the Facility Manager

Know your business	Submit an annual report for the department
Know and be able to use the language of business	Implement strategic facilities business planning
Understand the costs of doing business	Be able to develop, execute, and evaluate budgets
Become a skilled business communicator	Be a skilled contracting officer and procurer of goods and services
Identify and use best practices in all functions of facility management	Understand how you should manage, track, and report the ongoing performance metrics, stated in financial terms for the success of your department and service providers
Focus on cost reduction and on management improvements that will lead to cost reduction and cost avoidance	Think of ways to make well-run facilities a corporate advantage where appropriate
Understand, in detail, how you affect the business. Be able to translate facility management (FM) needs into FM requirements and to show how FM achievements fit business needs	For major decisions, use life-cycle costing
	Implement a regular program to communicate these metrics and your success to management and to your customers
Make your annual budget your principal facility management information tool	Understand depreciation and its effects on your budgets
Sign favorable leases and get control of your leases	Expect to invest in business technologies
In your practice and in your communications, stress the importance and benefits of good facility management	Understand the importance of being able to project and work to a budget and a schedule
Be able to use capital budget evaluation tools	Understand ratio analysis
Actively manage your real estate portfolio	Be able to administer chargebacks and allocations
Be capable of making lease-versus-buy decisions	Reduce churn

SOURCE: D. Cotts and E.P. Rondeau, 2004.

fectively with stakeholders and decision makers with differing technical backgrounds and at all management levels. Training of existing staff and the recruitment of new staff with such skills is required.

Academic institutions are developing facilities and infrastructure management programs and courses to educate both students and practitioners on approaches and methods for managing facilities and infrastructure as assets. Cornell

University, Eastern Michigan University, Ferris State University (Michigan), the University of Southern Colorado, and Brigham Young University all have programs in facility management (IFMA, 2003). George Mason University offers a professional certificate in facility management and the Georgia Institute of Technology offers a master's program in building construction and integrated facility management. Thirteen of 51 civil engineering and related programs surveyed had at least one course in or related to civil infrastructure management (Amekudzi et al., 2001). These developments point to a growing demand for formally trained facilities and infrastructure managers with both the technical expertise and business acumen to successfully manage facility portfolios and civil infrastructure systems as assets and to the many resources available nationwide that offer full- or part-time training.

EXAMPLES OF FACILITIES ASSET MANAGEMENT SYSTEMS

Included below are two examples of facilities asset management systems in use. The study committee did not evaluate their effectiveness, and their inclusion should not be viewed as an endorsement. However, the examples are indicative of several directions being taken.

The first example of a facilities asset management system is found at Brigham Young University (BYU). Implementation of BYU's asset management system began in the early 1980s, after the existing system had resulted in a culture of competition, confusion, and lack of trust among the various stakeholders. The search for a new system resulted in a paradigm shift from being a money-driven system to a requirements-driven system (Campbell, 2000).

The database developed to support this new paradigm tracks requirements (Figure 2.5), ensures that all assets are included in the inventory, and updates the assets based on life-cycle costing. Beyond standard maintenance and repair, an annual inspection is made of all assets that have 1 year of remaining life to assess whether their life can be extended or if replacement is warranted. Customer requests, one-time projects, and areas experiencing continual maintenance problems are also reviewed.

This system has led to a clearer understanding and definition of operating and capital budgets. Operating budgets include operations, maintenance, and repairs and must ensure that assets function and are managed properly, users are satisfied, and the environment is stable. Capital budgets include replacements, retrofits, improvements, and new space additions that maximize or extend the useful lives of facilities (Figure 2.6).

A partnership has been forged with all key stakeholders wherein an annual funding limit, a 40-year cash flow average of all life-cycle database items, and a 5-year average of all facility master plan items are mutually created. The annual funding limit in each area is reviewed periodically for required changes. Annual inspections and reviews are done to determine requirements. If the requirements

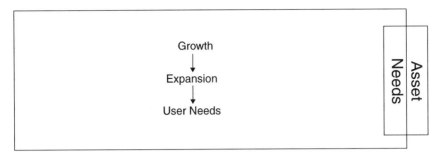

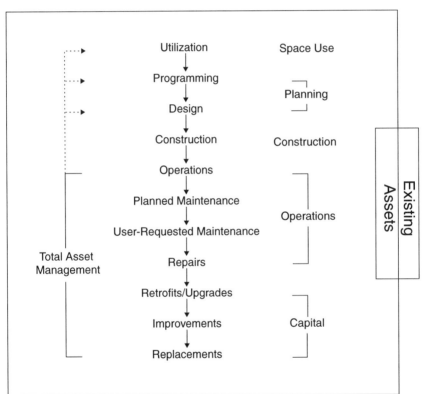

FIGURE 2.5 A facilities asset management structure (BYU).

do not exceed the limit, the difference goes into the "bank" for future use on that asset. If requirements exceed the annual fund limit, then those funds come out of the bank.

A second example of a facilities asset management system is being implemented at the University of North Carolina (UNC). UNC has a repairs and reno-

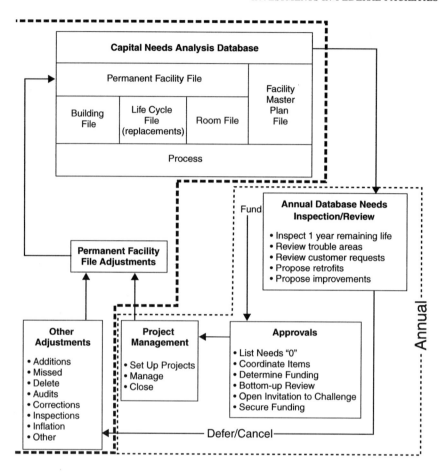

FIGURE 2.6 A facilities asset management framework (BYU).

vations reserve fund that provides an annual allocation for repairs and maintenance. To ensure that these funds are effectively spent, the university has developed a method for measuring the cost of work needed to bring a facility up to some baseline level of quality. This incorporates data kept by UNC's Facilities Condition Assessment Program (FCSP) and also goes beyond it. The FCSP identifies only the work required to bring a facility back to its original condition, as well as to correct life safety code deficiencies, while the recently developed Facility Condition and Quality Index (FCQI) also measures the cost to address functional and qualitative obsolescence relative to a desired baseline. This index divides the amount it would cost to bring a facility up to the desired functional level over the replacement value of that facility. For example, if a facility has a replacement value of $25 million and a cost of $2.5 million to bring it to the desired performance level, the FCQI would be 0.1 (Klein et al., 2002). An FCQI exceeding unity indicates that it would cost more to upgrade and modernize the facility in question than it would to build a new one. Where this occurs, the university automatically substitutes a replacement building into the 10-year capital needs plan. In such a case the existing building is not necessarily torn down but might be modernized to meet less demanding requirements.

To arrive at the FCQI, UNC has a uniform method of compiling facilities condition data using an online questionnaire about the characteristics of each building (e.g., structural condition, accessibility, maintainability). The database that results is also maintained and manipulated online. Beyond quality data, project implementation data are also entered and tracked via the Web, with access available to relevant stakeholders.

PRINCIPLES AND POLICIES FROM BEST-PRACTICE ORGANIZATIONS

Based on a consolidation of research, interviews, briefings, and the committee members' individual and collective experience, the committee found that best-practice organizations operate under a number of principles and policies (all 10 principles/policies are repeated in Chapter 6). In matters of facilities management,

> **Principle/Policy. Best-practice organizations implement a systematic facilities asset management approach that allows for a broad-based understanding of the condition and functionality of their facilities portfolios—as distinct from their individual projects—in relation to their organizational missions. Best-practice organizations ensure that their facilities and infrastructure managers possess both the technical expertise and the financial analysis skills to implement a portfolio-based approach.**

Facilities asset management is an evolving approach that helps to ensure that an organization's facilities portfolio is aligned with its mission. Required elements include accurate data about the facilities' portfolio; models for predicting the future condition of these facilities and the performance obtainable from them; engineering and economic decision support tools for trade-off analyses among competing investment alternatives; performance measures to evaluate the impacts of different types of actions (e.g., maintenance versus rehabilitation) and the timing of investments on the overall goals for facility provision; and short- and long-term feedback procedures.

Implementation of a facilities asset management approach requires facilities and infrastructure managers with the technical expertise found in traditional facilities management organizations (e.g., engineering, architecture, mechanical, electrical, contracting) as well as an understanding of financial concepts and management.

3

Decision Making to Support Organizational Missions

BACKGROUND

Organizations are established to achieve specific goals and missions. Their level of success depends, in large part, on the effectiveness of their decision making. Every decision made by an organization is intended to make something happen that otherwise would not or to prevent something from happening that otherwise would (Ackoff, 1999).

Because of the sums of money involved and the long-term nature of facilities, major facilities investment decisions have direct impacts on many business units, operating groups, and management levels, as well as on the financial prospects of any large organization. Thus, multiple internal and external stakeholders are either directly or indirectly involved in and impacted by these decisions. These stakeholders typically have differing, and possibly conflicting, objectives, responsibilities, and levels of technical knowledge.

The magnitude of the financial resources required for facilities investments precludes investment in other activities of importance and thus requires explicit trade-offs—if x million dollars are invested in facility A as requested by stakeholders 1, 2, and 3, then x million dollars will not be invested in activities B, C, and D, as requested by stakeholders 4, 5, and 6. The potential for adversarial relationships, miscommunication, and gamesmanship among the stakeholders is obvious as each group seeks to achieve its own goals and objectives.

To help align the objectives, goals, and values of the various stakeholders toward achieving the organization's goals and missions, best-practice organizations establish a framework of procedures, required information, and valuation criteria to support their decision making about facilities requirements. The various components of the framework are understood and used by all relevant leadership, management levels, and operating groups, which helps to permeate a facilities asset management approach into the culture of the organization.

For facilities investment decisions, the components of the framework include:

- Common terminology,
- A basis of shared information,
- Decision processes that are clearly defined and incorporate multiple decision points,
- Performance measures,
- Feedback processes,
- Methods for establishing accountability, and
- Incentives for groups and individuals.

Together these components support decision making related to facilities requirements and investments, create an effective decision-making environment, and provide a basis for measuring and improving facilities investment outcomes.

This chapter features those components of a framework related to facilities requirements and investments. The roles of technical analysis and values in decision making are first reviewed. The following sections discuss management approaches to achieving a mission; information for decision making; and decision-making processes. Chapter 3 concludes with a summary of principles and policies from best-practice organizations.

THE ROLES OF ANALYSIS AND VALUES IN DECISION MAKING

There is a generally recognized five-step process to help guide decisions on issues worthy of careful thought (Hammond et al., 1999):

1. Define the decision problem.
2. Specify appropriate objectives.
3. Identify a full range of alternatives for meeting the objectives.
4. Understand the consequences of the competing alternatives.
5. Evaluate the alternatives, incorporating the necessary trade-offs.

Regardless of who owns or manages them, facilities are built or renovated as a result of a similar decision process:

- The requirement for a facility to serve a specific function or purpose is identified.
- A set of objectives is developed for the facility.
- Different alternatives for meeting the objectives are identified.
- The consequences of the alternatives are estimated.

- Trade-offs are made to evaluate the alternatives.
- A decision is made to proceed.

A number of activities are then required to implement the program and to operate a facility. Many of these activities also occur as a result of decision processes:

- Funding is obtained.
- The facility is acquired through construction, renovation, lease, or purchase.
- The facility is occupied, operated, and maintained over a period of years and sometimes renewed.
- At the end of its life, the facility is disposed of.

Such processes appear logical and straightforward. However, in the real-life operating environments of corporations or federal agencies, where multiple stakeholder groups have a direct interest in the outcome of facilities investment decisions, decision making is rarely perfectly logical or sequential. Instead, decision making is likely to be interactive and iterative and to involve various stakeholder groups, who have different interests and information, at different and multiple points in the process.

Furthermore, decisions of any import are not based solely on technical analysis. The various parties involved also judge the desirability of the outcomes of various alternatives based on their individual and organizational values—that is, what an individual, a society, or an organization aspires to achieve: the health of human beings, the preservation of an ecosystem, an improved quality of life, or the ability to carry on an economic activity. When making decisions about investment alternatives, the various stakeholder groups use their values explicitly or implicitly to answer such questions as, How much of one service type should be given up to enhance another service type? How much is it worth to enhance the service quality of each type of service? Ultimately, values are at the core of all investment decisions and characterize the desirability of their consequences.

For large organizations, data, logical analysis, and judgments about facts help to determine the likelihood of the consequences of an alternative. Quantitative analysis can help people to systematically assess the implications of information and expose biases and flaws in their reasoning (Lempert et al., 2003). However, the decision-making process can quickly result in gridlock if the various stakeholders cannot agree on the assumptions that will form the basis of the analysis.

A further complication is that the desirability of the consequences will be judged differently by the different stakeholder groups based on their values. To understand how and why organizations make decisions, both types of judgments are important and must be accounted for. Confusing fact-based judgments with value-based judgments can lead to miscommunication, mistrust, and a decision-making environment characterized by adversarial relationships and gamesmanship (Kleindorfer et al., 1993).

To help align the values, goals, and objectives of the various stakeholders, an overarching desired outcome, such as mission achievement, must first be identified. The components of a framework to support achievement of that outcome can then be developed. For example, in the justice system, one overarching desired outcome is that anyone accused of a crime receive a fair trial. A jury of peers is assembled to decide on guilt or innocence. The prosecution and the defense, who have diametrically opposed objectives, work within a framework of procedures, required information, and valuation criteria to present their cases. They use a common basis of information or set of facts to build their cases, although they are free to reach differing conclusions. The information is deemed to be credible because it is provided under oath and penalties exist for perjury. The performance of the prosecution and the defense is measured by their success in swaying the jury to their point of view. The various arguments are tempered by a judge, who is responsible for ensuring that the appropriate procedures are followed to achieve a fair trial.

Best-practice organizations similarly establish a framework of procedures, required information, and valuation criteria to meet an overarching desired goal—achievement of mission. As noted in Chapter 2, a facilities asset management approach allows an organization to integrate facilities considerations into its strategic planning processes and to forge a direct link between organizational goals, investment decisions, and operations. The next section describes some management approaches that can be used to reinforce strategic decision making.

MANAGEMENT APPROACHES FOR ACHIEVING A MISSION

Best-practice organizations use their mission as guidance for instituting management approaches that integrate all of their resources—personnel (human capital), physical capital (facilities, inventories, vehicles, and equipment), financial capital, technologies, and information—in pursuit of a common goal. Ackoff describes two types of management approaches. The first, preactive planning, is a top-down, strategically oriented approach based on forecasts of suppliers, consumers, and competitive behavior as well as economic, social, and political conditions for which senior management sets organizational objectives. The tactics for meeting these objectives are left to the individual operating units. The second approach, interactive planning, is directed at gaining control of the future and consists of the "design of a desirable future and the selection or invention of ways of bringing it about as closely as possible." Interactive planning focuses on involving personnel from within the organization in the planning process so that they can "come to understand their organization and its environment, and how their behavior can improve performance of the whole, not just their part of it" (Ackoff, 1999, p. 106).

Yet another management approach for integrating the use of resources is one that focuses on an organization's essential areas of expertise (its core competencies), which are the organizational skills that are difficult to duplicate, that create a unique value, or that constitute the organization's competitive advantage—that is, what it does better than anyone else (NRC, 2000).

In this approach, functions deemed to be core competencies are assigned to an organization's in-house staff because they have the skills and institutional knowledge to most effectively perform them. In-house staff may also perform functions that support core competencies to keep competitors from learning, taking over, eroding, or bypassing the organization's core competencies (Pint and Baldwin, 1997). Noncore functions that are required but not critical to an organization's competitive position—for example, janitorial services—may be outsourced to providers with expertise in that function.

Using this management approach, facilities investments can be evaluated based on their support of the organization's mission and core competencies. For example, if the core competencies are research and development of new pharmaceutical products, then laboratories and other research or manufacturing facilities can be directly linked to operations essential to the organizational mission and evaluated as mission enablers. Facilities that support core competencies—for example, administrative space required for in-house staff or noncore functions—can be differentiated from facilities viewed as mission enablers.

Level of Control and Planning Horizons

When considering a facilities investment proposal, best-practice organizations determine the level of control required and the planning horizon (the length of time a facility will be needed to support a particular function), which may or may not be the same as the life of the facility.

Based in part on the level of control an organization wishes to exert over its facilities, it may choose to own them or lease them. Ownership allows the organization to exert maximum control over a facility's condition, functionality, and operations. In choosing ownership, an organization takes a risk that if requirements change, the facility can be disposed of without a substantial loss. It also takes on a financial commitment to operate and maintain the facility over time. However, the owner can realize financial benefits if opportunities arise to sell a property at a profit. If a facility is demolished, the owner may be able to realize some salvage value.

By leasing space,[1] an organization gives up some control: for example, the

---

[1] The option of leasing facilities presumes that such facilities are available in the marketplace.

lessor's approval might be needed for any modifications, or the term of the lease might affect the organization's ability to reduce costs by moving out. The lessor could also choose not to renew a lease or to offer to renew it only at a higher rate.

The advantages of leasing include lower up-front capital and financing costs and less restrictive credit standards, which translate into less risk and greater liquidity (how easily assets can be converted to cash). An organization can choose to renew the lease periodically, allowing it to adjust its space needs to reflect evolving operational requirements. If the space becomes obsolete, is no longer required, or is in the wrong location to best support current operations, the organization can move elsewhere, leaving the lessor to pay the costs of ownership and obsolescence.

The type of lease entered into (operating or capital[2]) will depend on the type of function to be supported, the organization's financial position, its desire for flexibility, and its operating environment. Whatever the type of lease, an organization cannot claim any tax depreciation benefits or realize any residual values through sale or salvage value through demolition.

The General Motors Corporation illustrates one way among many of how a facilities asset management approach can be directly linked to organizational mission and strategic planning. General Motors (GM) has identified its manufacturing plants as directly supportive of its core competencies and operating requirements—designing and producing vehicles. GM exerts maximum control over these specialized facilities by owning them for an indefinite period of time and staffing them with its own workforce.

GM has also developed a strategy for nonmanufacturing facilities intended to provide a scalable portfolio that responds to changing business needs (GM, 2003). To leverage facilities investments, nonmanufacturing facilities have been divided into three investment and use classifications (see Table 3.1). "Committed" facilities involve a long-term commitment. They are owned by the corporation to allow for proprietary investments and to be used primarily by the corporation's internal staff, although contractors, suppliers, or alliance partners that support the corporation's core business may occupy some of this space. A second category is "flex facilities," which are mid-term investments that allow GM to exit from the space relatively rapidly if requirements change. Because flex facilities are designed to accommodate a range of functions and appeal to a wider audience, they can be more easily disposed of in the marketplace. These facilities are owned and used by internal and noncorporation tenants. As demand changes, the amount of space devoted to flex facilities can be increased or decreased to balance the portfolio. The third category is "buffer" facilities, which support

---

[2]An operating lease is a lease usually lasting for 5 years or less in which the lessor handles maintenance and servicing. It may be most appropriate for short-term needs or in unstable markets. Capital leases, in contrast, are long-term leases, usually 6 years or more (Groppelli and Nikbakht, 2000).

TABLE 3.1 An Approach for Nonmanufacturing Facilities (GM)

Facility Category	Planning Horizon	Level of Control	Tenancy
Committed	Indefinite	Own	Internal staff
Flex	Mid-range	Own	Internal staff/contractors
Buffer	Short term	Lease	Tenants

noncore functions. Buffer facilities are used as space for tenants, have short-term leases, and can be easily disposed of in response to short-term business fluctuations.

INFORMATION FOR DECISION MAKING

To provide a basis for informed decision making about facilities investments, best-practice organizations foster communication among the various stakeholder groups through the use of common terminology; rigorously analyze and evaluate facilities investment proposals; and analyze ways to disengage from the proposed investment (exit strategies).

Common Terminology

Facilities investments typically are of a magnitude that can affect an organization's financial health: Decisions about whether to invest will impact many operating units. As noted in Chapter 1, private-sector organizations typically make capital investment decisions separately from decisions regarding operating expenditures. Best-practice organizations use a decision-making process for capital expenditures that involves many of the operating units at some point. However, engineers, accountants, facilities managers, senior executives, finance and tax experts, and market, technology, and personnel specialists lack a common vocabulary or style of interaction. Lack of a common terminology can easily lead to miscommunication about potential facilities investments and time delays that can have financial impacts.

Consider the concept "facility life." Building service life has been defined as the period of time over which a building, component, or subsystem provides adequate performance (NRC, 1991). Design service life is the time period building owners, designers, and managers use to make decisions about maintenance, repairs, operations, and alterations, typically between 10 and 30 years (NRC, 1990). Life cycle has been defined as the sequence of events in planning, design, construction, use, and disposal (e.g., through sale, demolition, or substantial renovation) during the economic or service life of a facility; it may include changes in use and reconstruction (NRC, 1991). Unless such terms are clearly defined and consistently used by all of the individual stakeholders, the potential for miscommunication is evident.

To communicate effectively across the various operating units—facilities, administration, finance, human resources, and marketing, among others—best-practice organizations establish and consistently use an agreed-upon set of terms to promote mutual understanding of the issues, risks, and possible outcomes of an investment proposal. Terms such as "capital" are clearly defined for use by all operating units in both proposal documentation and in interactive discussions so that time is not lost through miscommunication or by continually redefining the ground rules.

Business Case Analysis

To further enhance communication among the various stakeholders and to facilitate effective decision making, best-practice organizations use a business case analysis. A business case analysis is a tool for planning and decision making that projects the financial implications and other organizational consequences of a proposed action (Schmidt, 2003a). It links estimates of costs and benefits with expectations for projected outcomes. Although at its heart the business case is a financial analysis, it also contains information on organizational impacts that cannot be quantified in monetary terms, such as mission-readiness or fulfillment, customer satisfaction, and public image.

The overriding purpose of a business case analysis is to make transparent to the various decision-making and operating groups all of the objectives to be met by a facilities investment, the underlying assumptions, and the attendant costs and potential consequences of alternative actions. All of the participating groups in a facilities investment decision use the same analysis and its various refinements.

For these reasons, a business case analysis is designed and developed to answer questions such as, What are the likely financial and other business consequences if the organization takes a particular action? Which alternative for action represents the best business decision? Will the returns justify the investment? What will this action do for overall organizational performance? (Schmidt, 2003a, 2003b). Thus, a business case analysis is a planning and decision support tool, not a budget, an accounting document, or a financial reporting statement. Best-practice organizations treat a business case analysis as a living tool, one that is being continually revisited, refined, and updated, not as a static, one-time-only case study.

The format and types of analyses included in a business case analysis are not standardized: each organization determines and reaches general agreement on the types of data, analyses, and methodologies to be used and how that information will be presented. These components are strengthened over time through repeated use. The credibility and value of the analyses and methodologies are improved by understanding the types of information that are useful in differentiating the consequences of various alternatives.

Financial and other quantifiable objectives, together with objectives that are difficult to place a dollar value on, such as improved employee morale or improved corporate image, are identified up front. Because some assumptions and data underlying a proposal will be subjective and time sensitive (e.g., interest rates), the sources of all information related to business trends, future interest rates, inflation, salaries, and the like are documented. To provide credibility and accountability, the persons or business units that developed the proposal are identified (Schmidt, 2003b).

Best-practice organizations recognize the interrelationships among their people, places, other physical assets, technologies, information, and funds: A change in the character, size, or amount of any one of these resources will impact the other resources and the organization's ability to meet its goals and mission. In a business case analysis, such organizations analyze the life-cycle costs of a specific facility investment proposal and of the attendant staffing and equipment, and they look at alternative uses of the required funding over the appropriate planning horizon. They include the costs to finance the investment, the potential costs and benefits of disposal, including sale and salvage value, the costs of technology, and operational requirements. These analyses allow decision makers to better understand the potential consequences of facilities investment decisions and to make informed choices in regard to owning, leasing, reinvesting in, or constructing facilities.

Pro Forma Statement

At the heart of the business case is a pro forma statement that is essentially a financial analysis. A number of standardized, repeatable, analytical measures are typically used. These include net present value, internal rate of return, discounted cash flow, return on equity, return on net assets, and earnings per share.[3] The metrics chosen are those that best represent the values of the organization. Once developed, these metrics can be used to determine the cost of ownership, the benefit/cost ratio, or the cost-effectiveness index—all important decision-making criteria.

The financial information and assumptions used to develop the business case analysis must be carefully explained and documented because, owing to the compounding (or its reciprocal, discounting) effect of interest rates, all of the

---

[3]Return on investment is not a standardized, analytical measure; instead it is a concept whose definition varies by organization and discipline. Organizations using the term "return on investment" must clearly define how it is being used and how it is being calculated (Schmidt, 2003a).

financial metrics mentioned above are highly dependent on time and the cost of borrowing. For example, if the prevailing interest rate is 3 per cent, then a dollar either received or expended 5 years in the future is worth only $0.78 today (its "present value"), and a business case analysis must be careful to express all monetary costs and benefits in similar terms. As interest rates rise or the period of analysis lengthens, the present value of future costs or benefits decreases sharply. For example, if interest rates are 8 per cent, the present value of a dollar received or expended in 20 years is only $0.21. Even though the objectives of capital investment in the public sector differ from those of the private sector, the impact of time and interest rates on public-sector investment decisions is equally powerful.

Several types of financial analyses can be used to evaluate a particular action. Three with applicability to the public sector will be discussed here: cost of building ownership, benefit/cost, and cost-effectiveness.

Cost of Building Ownership

The cost of ownership of a building has been defined as the total of all expenditures an owner will make over the course of the building's service lifetime (NRC, 1990). The cost of ownership typically will include planning, design, and construction (first costs); maintenance, repairs, replacements, and alterations; normal operations such as heating, cooling, and lighting; and disposal. These costs are also referred to as life-cycle costs.

Benefit/Cost Analysis

A common method of selecting among alternative investments is to determine the ratio of a project's total benefits to its total costs—that is, the benefit/cost ratio. A benefit/cost ratio greater than 1.0 indicates that the benefits of the project outweigh the costs, while a ratio less than 1.0 means the opposite. Obviously, the higher the ratio for a particular alternative, the more attractive that project will be relative to other competing projects. Using benefit/cost analysis requires considerable care because the costs and benefits will be experienced at different times and their magnitudes may vary considerably. For example, in a building project, the relatively large first costs will be experienced early in the project's life and followed by smaller recurrent costs for maintenance, operations, repair, and replacement. Benefits generally will be small or nonexistent initially but may accrue to fairly large values late in the life of the project. Although the costs and benefits can be discounted to a single present value, doing so will require multiple assumptions about interest rates, timing, and the future values of these elements. Despite these cautions, benefit/cost analysis can be a powerful tool for evaluating alternatives.

Cost-Effectiveness Analysis

Benefit/cost analysis is predicated on the ability to express benefits in monetary terms, either as a cash inflow or a cost avoided. However, when only the cost side of a project can be quantified (as is often the case in public capital investment decisions), an alternative means of analysis and comparison is required. Cost-effectiveness analysis was developed as a means of evaluating environmental projects where, for example, the benefits of enhanced air or water quality or the value of wetlands were difficult or impossible to quantify accurately in monetary terms. In these cases, performance objectives were established for the action, and the project that met all desired objectives at the lowest cost was considered the most cost-effective. Despite mixed success with efforts to monetize environmental benefits, cost-effectiveness analysis is a useful business case tool when only the costs of a project are well defined.

Exit Strategies

To provide important insight about the potential consequences of investing in a long-term, nonliquid asset like a facility and to select the best alternative to meet the requirement, best-practice organizations typically develop and evaluate exit strategies—methods for disengaging from an investment—as part of the business case analysis.

A commonly analyzed and implemented exit strategy is to lease the required space in the first place. If requirements change, an organization can move out of leased space relatively quickly without the burden of selling or otherwise disposing of the property. In some cases, leased space may have a higher annual cost per square foot than owned space. However, it may still make economic sense to lease to ensure that the organization can divest itself of the space on short notice.

For space that is to be acquired through purchase or construction, one exit strategy is to build flexible (generic) space that can be relatively easily adapted to other uses to meet changing requirements. Flexible office or warehouse space generally has wider appeal to potential buyers or those willing to sublease excess space; this can mitigate the risk of selling it at a financial loss and increase opportunities for selling it at a profit. Johnson and Johnson, for example, builds its biopharmaceutical facilities using flexible floor plans. With rapidly changing markets and an 8-year-long Food and Drug Administration approval process, the risk is considerable that when a project is completed, it may be outmoded or its intended product lines will not gain approval. Johnson and Johnson mitigates the risk by constructing facilities that can be relatively easily adapted to new uses or different product lines. The Toyota Corporation takes a different approach, building in flexibility by constructing large facilities that are similar to one another in order to accommodate a broad range of uses and to reduce surprises—a portfolio approach.

Timely maintenance and repair of an owned facility can also be evaluated as an exit strategy: Investment in maintenance and repair retains or improves the functionality and performance of a facility, thereby increasing its marketability and its residual value at the time of sale.

As the merits of a proposal are evaluated, the costs and benefits of leasing versus owning, of developing flexible facilities, and of maintenance and repair, as well as the projected residual value, are analyzed to provide quality information for decision making. Tishman Speyer Properties, for example, develops and evaluates at least two exit strategies for every proposed investment.

For some specialized facilities, such as those for manufacturing, power generation, defense or military use, and some types of research, the only exit strategy may be demolition, cleanup, and disposal. A particularly strong rationale is needed for investing in such facilities, such as a direct link to the core business lines and missions of an organization, and the cost of the intended exit strategy must be made explicit in the initial proposal. This exit strategy is evaluated to provide information about the total costs involved and to provide insight into design and operation practices that may lead to lower demolition and cleanup costs. For example, the use of biodegradable materials for a facility may result in lower disposal costs, or special waste disposal methods may be indicated.

DECISION-MAKING PROCESSES

In private-sector organizations, decisions about facilities investments are typically made by a senior executive-level group—an investment committee, a management committee, a group of senior vice presidents representing all of the operating units, or the board of directors. This decision-making group is responsible for ensuring that facilities investments are integrated into the overall organizational strategy. The decision-making body reviews a proposal at several stages of development. Each stage represents a decision point at which the reviewing body will decide if the proposal should be given conditional approval and considered further or if it should be terminated (go/no-go determination).

Funding thresholds are established to determine the level at which a proposal will be reviewed—the greater the cost or potential impact, the higher the level of management review. The board of directors may make the final decision about investment proposals with potentially significant impacts on the organization's cash flow, productivity, or competitiveness; in this case, an executive-level reviewing body will forward the proposal to the board as a recommendation rather than a decision.

Minimal resources are invested at the earliest stages of proposal evaluation, and the business case analysis is likely to focus on the financial aspects of the proposal, the pro forma statement. As a proposal receives conditional approvals, and as additional resources are committed, more detailed analyses are undertaken, and the business case documentation becomes more complete until the proposal becomes an actual project. Once final approval for a project is received, it is usually put on a fast track so that the resulting facility can be functional as soon as possible.

Throughout the process, information is continually gathered, refined, documented, and updated. Decisions are continually revisited to determine if modifications are needed in response to changing requirements. All significant decisions are documented to create a decision record that can be archived and revisited. Such a record creates an institutional memory and allows the organization to save time when reevaluating a decision and when orienting people to the project as leadership and managers change.

Figure 3.1 depicts a typical process for facilities investment decision making used in best-practice organizations. The following text elaborates on individual elements of this process.

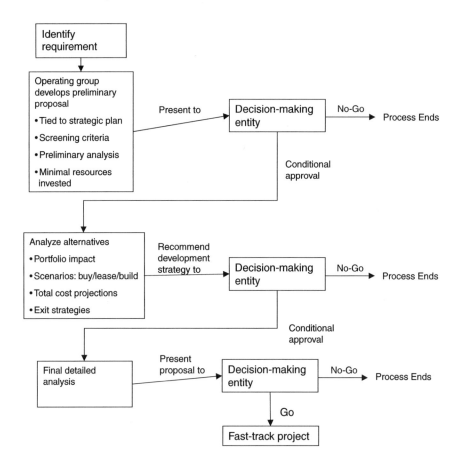

FIGURE 3.1 Typical decision-making process for facilities investments.

Identifying a Facility Requirement

In a best-practice organization, a proposal involving a facility investment may come from any of the operating units within the organization. The proposal must be tied to the organization's missions, organizational objectives, long-term or rolling capital plan, and sometimes to an individual business unit's annual plan and goals. It also must meet established screening criteria (e.g., opportunities to make money, avoid costs, improve customer satisfaction, improve product delivery, or create operating efficiencies).

Typically, a facility investment proposal is presented as an opportunity for the organization to make or save money, avoid costs, or comply with regulations. Opportunities for making money might occur where there is a facility requirement tied to an increased demand for a good or service attributable to increases in population, increases in income, or an influx of new businesses. Opportunities for saving money might be realized by creating operating efficiencies, by making improvements that minimize the potential for accidents or other liability actions, by replacing an obsolete facility with one that is state of the art, by consolidating facilities, or by disposing of facilities that are no longer needed. Costs might be avoided by incorporating nontoxic or recyclable materials in a building to avoid the additional expense of disposing of hazardous materials at demolition. Or, an investment might be proposed to comply with regulatory requirements, local building codes, environmental standards, or new mandates.

At this initial stage, the level of analysis must be sufficient to determine whether the proposal has merit, without incurring significant time and resources. A pro forma statement might include the underlying assumptions, preliminary estimates of internal rate of return, cash availability (expected costs and cash flow), ledger impact (depreciable expense), and asset burden (tax flow), as well as judgments about the potential impact on the organization's operations, market risk, and opportunities. In private-sector organizations, an earnings per share analysis might be included to demonstrate the impact on profits and earnings. At this stage, the information presented is high level and succinct, and the pro forma may include some "plug in" numbers. The business case analysis may be limited to the vision, the opportunity, the long-term benefit, a plan, and a net present value analysis comparing the life-cycle cost of a lease with the life-cycle cost of owning a facility.

An investment proposal is presented to the reviewing body by its organizational "owner," typically the head of an operating unit that has a stake in its successful outcome. The reviewing body will decide if the proposal has merit and should be conditionally approved pending additional analysis or if it should be terminated (no-go).

Recommending a Development Strategy

If conditional approval is given, more detailed analyses are undertaken as the business case is developed. Typically, however, a wide range of alternatives for meeting the requirement will be developed, including an alternative for not making a facility investment. The organization will also analyze how it can fulfill the requirement by squeezing production capacity out of the existing portfolio of facilities or meeting it through other, nonfacility alternatives, such as outsourcing. It will identify facilities in the portfolio that might become obsolete to the mission, underutilized, or overutilized if the proposal is implemented. If these analyses indicate that additional facilities are required, alternatives for buying, leasing, or building them and for disposing of facilities that are no longer required will be evaluated. The life-cycle costs of all required resources (operating, staffing, information technologies, financial, facilities) are projected for each alternative.

What-if scenarios or sensitivity analyses that change the assumptions about a proposal are used to aid in deliberation and decision making. Scenario development and evaluation can identify a range of situations that are sufficiently plausible and then evaluate their relative risks, costs, and benefits related to cash flow, profits, life safety (e.g., accidents, injury, fire, earthquakes) and security, environmental impacts, and the like. If the original proposal does not meet the investment objectives, its scope may be changed to consider the effects of a lower-cost alternative.

All of this information is returned to the appropriate reviewing body. The level of information presented must be sufficient for all the decision makers to understand the trade-offs involved in choosing one alternative over another. At this decision point, the reviewing body may narrow down the alternatives, request more analysis, or terminate the proposal.

A final, detailed business case analysis is then completed. The required information may be prepared by cross-functional teams, individual business units, contractors, or some combination of these, depending on the culture and resources of the organization. The numbers are validated by the various operating units, including the facilities management group. In some cases, an independent third party may be hired to verify the numbers. Based on this information, a development strategy is recommended. The proposal is again taken to the reviewing body for a go/no-go decision.

Time Frame and Continuous Evaluation

On paper, such a process appears to be lengthy and time consuming. In practice, executive-level committees of private-sector corporations meet as often as once a week. Even if a proposal goes to a reviewing body four or more separate times, it may take less than 6 months to move from initial review to final approval. It is not uncommon for a proposal to go from the planning process to occupancy of the resulting facility in less than 3 years. In many cases, projects are linked to the production schedule of a new product or service, so the timeline is set by the schedule for production or service availability.

Project implementation may be delayed if there is a change in the external operating environment, such as a change in interest rates, if a tenant must be committed to a project before construction begins, if a rezoning approval is needed, or if difficulties arise in bringing a contractor on board. If the operating environment changes substantially or significant time elapses before the project can be initiated, best-practice organizations reevaluate the decision to approve the project and determine whether to proceed or cancel it.

PRINCIPLES AND POLICIES FROM BEST-PRACTICE ORGANIZATIONS

Based on a consolidation of research, interviews, briefings, and the committee members' individual and collective experience, the committee found that best-practice organizations that successfully manage facilities investments operate under a number of principles and policies in their decision making (all 10 principles/policies are repeated in Chapter 6):

Principle/Policy. Best-practice organizations establish a framework of procedures, required information, and valuation criteria that aligns the goals, objectives, and values of their individual decision-making and operating groups to achieve the organization's overall mission. The components of the framework are understood and used by all leadership and management levels.[4]

In large organizations, significant facilities investment decisions typically entail millions of dollars and have direct impacts on many divisions, operating groups, management levels, and budgeting processes. Multiple internal and external stakeholders with differing objectives, responsibilities, and levels of technical knowledge are impacted by these decisions and the trade-offs required.

To align the values and objectives of all relevant decision-making and operating groups, best-practice organizations establish a framework of procedures, required information, and valuation criteria to support effective decision making. Components include common terminology, a business case analysis, and evaluation processes that are clearly defined and involve multiple decision points.

Principle/Policy. Best-practice organizations integrate facilities investment decisions into their organizational strategic planning processes. Best-practice organizations evaluate facilities investment proposals as mission enablers rather than solely as costs.

---

[4]This principle/policy and the principle/policy in the Executive Summary and at the end of Chapter 4 together form Principle/Policy 1 in Chapter 6.

Best-practice organizations institute decision-making and management approaches that integrate the use of all of their resources—people, financial, facilities and other physical assets, technologies, and information—in pursuit of mission achievement. They evaluate facilities investment proposals as mission enablers rather than solely as costs: Investments in facilities are typically made to ensure that business operations are continuous and efficient, essential ingredients to an organization's success. Executive-level managers from all of the operating units are responsible for reviewing facilities investment proposals, making decisions about their viability, and ensuring that facilities investments are integrated into the organization's overall strategic planning processes.

Principle/Policy. Best-practice organizations use business case analyses to rigorously evaluate major facilities investment proposals and to make transparent a proposal's underlying assumptions; the alternatives considered; a full range of costs and benefits; and the potential consequences for their organizations.

A business case analysis is a planning and decision-support tool used to ensure that the objectives for a proposed facility-related investment are clearly defined; a broad range of alternatives for meeting the objectives is developed; the alternatives are evaluated to determine how well the objectives will be met; and trade-offs are explicit. It is a living tool that is continually revisited, refined, and updated throughout the decision-making process.

Principle/Policy. Best-practice organizations analyze the life-cycle costs of proposed facilities, the life-cycle costs of staffing and equipment inherent to the proposal, and the life-cycle costs of the required funding.

Best-practice organizations recognize the interrelationships among their people, places, physical assets, technologies, information, and funds: A change in the character, size, or amount of any one of these resources will have impacts on the other resources and the organization's ability to achieve its mission. Within a business case analysis, best-practice organizations analyze the life-cycle costs of proposed facility investments in addition to the first costs (design and construction), the costs of financing the investment, the potential costs and benefits of disposal (sale and salvage value), and life-cycle costs and benefits related to staffing, technology, and operational requirements.

Principle/Policy. Best-practice organizations evaluate ways to disengage from or exit facilities investments as part of the business case analysis and include disposal costs in the facilities life-cycle cost to help select the best solution to meet the requirement.

Best-practice organizations typically consider how they can disengage from a proposed investment (exit strategy) at the same time they are determining whether or not to proceed with it. Commonly analyzed and implemented exit strategies include leasing rather than owning the required space; acquiring flexible or generic space that offers more options to the owner and that might appeal to a wide range of potential buyers; and timely maintenance and repair, which increase a facility's marketability and residual value at the time of sale. For those facilities where the only viable exit strategy is demolition, cleanup, and disposal, the costs of the activities are estimated for the business case analysis; these projected costs, in turn, may influence the eventual design of the facility, choice of materials, and methods of operation.

Principle/Policy. Best-practice organizations base decisions to own or lease facilities on the level of control required and the planning horizon for the function, which may or may not be the same as the life of the facility.

When considering a facilities investment proposal, best-practice organizations determine the level of control (own or lease) they wish to exert over facility conditions and operations based on the function's importance (core competency or noncore function). They also consider the planning horizon—the length of time the property will be required to support a particular function, which may or may not be the same as the life of the facility.

4

Environments for Effective Decision Making

BACKGROUND

Ultimately, of course, good decisions are made by good decision makers armed with credible information, insights from analysis, and appropriate skills. Information and processes are some of the ingredients needed. The environment within which decision makers employ these ingredients is also important because it will affect the free flow and interchange of information and conclusions.

As stated in Chapter 3, best-practice organizations establish a framework of processes, required information, and valuation criteria that aligns the individual goals, objectives, and values of its decision-making and operating groups so as to achieve the organization's mission. A framework also helps to create an effective decision-making environment and to provide a basis for measuring and improving the outcomes of facilities investments. This chapter first discusses the role of open communications, trust, and credible information in creating effective decision-making environments. The focus then shifts to the use of performance measures, continuous evaluation and feedback processes, accountability, and incentives. The chapter concludes with principles and policies from best-practice organizations.

OPEN COMMUNICATIONS, TRUST, AND CREDIBLE INFORMATION

Communication can be defined as "the science and practice of transmitting information, normally through the use of symbols, in a manner that succeeds in evoking understanding."[1] It is, therefore, more than a good presentation or a dynamic messenger. Communication is about the quality of the message, the credibility of the information, and the deliberations that ensue. Effective communication among individuals, business units, or a range of stakeholders can be difficult to achieve because there are many opportunities for distorting the message, information, and deliberations. Barriers to effective communications include lack of a common terminology; lack of trust in the source of information; poor interpersonal relationships; differing individual and group values; and unexpressed assumptions.

Terminology is a factor because different people often interpret the same words differently based on their professional training, experience, or values. If the source of the information is not credible, the believability of the overall message may be called into question. How people receive a message will depend, in part, on their past experiences with the person delivering the message as well as their relationships within and to the organization. Assumptions come into play when people take for granted that others see the situation in the same way and will have the same reaction.

In best-practice organizations, effective decision making for facilities investments is related to managing a free exchange of information among the various stakeholders, particularly those who might be skeptical about a proposed investment. Open communications ensure that those who need to know and who can best critique a proposal have access at a sufficiently early stage to provide information and insights that can be constructively used to produce a better proposal. The more open the process, the more likely it is that errors in fact or methodology will be uncovered.[2]

Trust—unquestioning belief in and reliance on someone or something—is widely understood to be important to the success of almost all forms of human interaction. Building trust is a complex prospect. Trust is fragile. It is difficult to establish and easy to destroy. One incident of poor communication could be interpreted as deceptive and trust can be lost. One realization that a game has been played wherein one player was less than open and honest can destroy trust. Numerous documented, positive transactions are required to build trust (Slovic, 1993).

Effective communication is both a top-down and a bottom-up responsibility. Senior executives are responsible for ensuring effective communication of policy decisions and institutional strategies throughout an organization. They must cre-

---

[1] Modified from definitions in *The New Shorter Oxford English Dictionary,* Oxford: Clarendon Press, 1993.

[2] Closed processes may be required in situations involving the protection of proprietary information, safeguarding national security, and judicial proceedings, among others. There is always a risk in open proceedings, especially in the public sector, that a multitude of participants with different motivations may delay or sidetrack a process.

ate an environment that encourages a flow of communication from all management levels without fear of reprisal. Mid-level and line managers lack the position in the organization to bring all of the relevant stakeholders together and move them in a common direction. Nonetheless, the managers of operating units are also responsible for communicating effectively. It is incumbent on the real estate manager to understand competitive strategy, not on the line manager to understand real estate (O'Mara, 1999).

As described in Chapter 3, common terminology and a business case analysis are used to foster effective communication among the various stakeholders in discussing proposals for facilities investments. The business case analysis is seen as credible because it is understood and used by all relevant leadership, management levels, and operating groups. Common terminology and a business case analysis also serve to create a basis for open communications and to build trust. The business case analysis provides a means to review the strategic, qualitative, and quantitative aspects of a proposal and compare it with other proposals. When the analysis is used in deliberations, discussion of the underlying assumptions allows everyone to see where everyone else stands on the proposal, to identify a full range of alternatives, and to discuss their merits and deficiencies in meeting organizational objectives, as opposed to operating unit or individual objectives. Expertise in developing and using analyses and communicating the results permeates the organization's workforce and culture when all participants use the same information repeatedly. Trust is built among the decision-making and operating groups by ensuring that everyone has access to the same information.

Persons interviewed for this study noted that facilities management operating groups had gained or retained credibility and built trust at the institutional level by providing sound information, by incorporating rigor into their analyses, by giving high-quality presentations, and by submitting realistic, reasonable requests for investment proposals. Among the specific examples cited were the following:

- Providing good cost estimates the first time around. The cost estimates were developed using cross-functional teams and reviewed by an internal cost estimator before being presented to the executive board (Dallas-Fort Worth airport).
- Having the division director who would be responsible for building and operating a proposed facility present the proposal to the board of directors. Having a presenter with operational responsibility and a successful track record increased the credibility of the presentation (Public Service Corporation of New Mexico).
- Including input from facilities operating and maintenance staff in the business case analysis (DMJM + Harris).
- Closing out projects and turning back unused contingency funds. As a

result, members of the facilities management operating group were seen by others in the organization as timely performers and good financial stewards. Over time, the group's budget was increased because it was trusted to use the funds wisely (Dallas-Fort Worth airport).

Decisions about proposals for facilities investments are linked to organizational mission and take into account the organization's portfolio of facilities. Such decisions, however, are made within a dynamic environment where operational requirements and even the mission of the organization are subject to change. Best-practice organizations use additional framework components—performance measures, feedback processes, accountability, and incentives—to measure, adjust, and improve decision-making processes, management practices, and the results or outcomes of decisions.

PERFORMANCE MEASURES

For any organization, it is important to understand why decision-making processes or management practices that had been used led to success or failure, and how that understanding can suggest improvements. The notion of continuous process monitoring and feedback is based on the recognition that, however effectively one plans, unintended consequences, unforeseen events, and change will occur. Best-practice organizations measure the results or outcomes of facilities investments by establishing baselines[3] and performance measures[4] to constantly monitor and track all aspects of operations and their results in relation to organizational objectives. Performance measures help to identify where objectives are not being met, or where they are being exceeded. Managers can then investigate the factors or reasons underlying the performance and make appropriate adjustments.

Best-practice organizations have long used metrics such as internal rate of return, growth or decline in earnings per share, percentage of market share, and the like to measure overall performance in relation to mission and the desired results. However, because such measures may focus on what has happened already—that is, on investments already made—they may not be particularly useful for planning for the future or responding to changing requirements. For those purposes, operational measures are required that focus on elements that are important to future financial performance, such as the level of customer satisfaction or the introduction of innovative products, techniques, or technologies. A study conducted in the early 1990s found that

> Executives also understand that traditional financial accounting measures like

---

[3]Defined as a quantifiable point at which an effort began and from which change can be measured and documented (NAPA, 1996).

[4]Defined as the standard by which to gauge an operation or activity (NAPA, 1996).

return-on-investment and earnings-per-share can give misleading signals for continuous improvement and innovation—activities today's competitive environment demands. The traditional financial performance measures worked well for the industrial era, but they are out of step with the skills and competencies companies are trying to master today.... Senior executives do not rely on one set of measures to the exclusion of the other. They realize that no single measure can provide a clear performance target or focus attention on the critical areas of the business. Managers want a balanced presentation of both financial and operational measures (Kaplan and Norton, 1992, p. 71).

This particular study led to the development of the Balanced Scorecard (BSC), a conceptual framework for evaluating organizational performance. Over time, the BSC has evolved, but the categories of performance to be measured have remained constant: financial outcomes, internal business processes, customer relationships, and innovation and learning.

"Balanced" refers to several qualities of the scorecard. First, there is a balance across the four categories to avoid overemphasis on financial outcomes. Second, it requires both quantitative and qualitative measures. Third, there is a balance in the levels of analysis—from individual and group results to organizational outcomes (Heerwagen, 2002). The BSC approach can be applied hierarchically, beginning with organizational objectives and cascading down to operating units and individuals with successively more detailed objectives and measures at lower levels.

Since its introduction, the BSC approach has been adopted and developed for use by many organizations, in the private, public, and not-for-profit sectors. Because it is a conceptual framework, each organization that implements a BSC approach must develop its own strategic objectives and the performance measures needed to evaluate progress toward those objectives.

For example, the Mobil North American Marketing and Refining Company identified return on capital used, existing asset utilization, profitability, and profit growth as financial strategic objectives; stakeholder objectives included customer satisfaction in each of the market segments it serves and good relations with its dealers; internal processes included objectives for the performance of facilities, inventory management, and on-time delivery of products of specified quality; learning and group objectives included improving core competencies and skills and providing employees access to the strategic information needed to do their jobs (Kaplan and Norton, 2001).

In the case of Charlotte, North Carolina, a public sector municipality, its strategic objectives for financial outcomes included increasing the tax base and maintaining the city's bond rating. Stakeholder relations comprised such objectives as increasing the perception of safety, enhancing service delivery, maintaining a competitive tax rate, and promoting economic opportunity. Strategic objectives for internal processes included improving productivity and increasing

TABLE 4.1 Strategic Assessment Model Matrix of the Association of Higher Education Facilities Officers (APPA)

FINANCIAL PERSPECTIVE	INTERNAL PERSPECTIVE
Facility Operating CRV Index	Cycle time
Facility Operating GSF Index	Average age
Capital Renewal Index	Backlog
Facilities Condition Index	Energy usage
Needs Index	Energy Reinvestment Index
CUSTOMER PERSPECTIVE	INNOVATION AND LEARNING PERSPECTIVE
Customer Satisfaction Index	Work Environment Index
Distribution Index	High Score Index
Weighting Index	Distribution Index
Gap analysis	Organizational change assessment

SOURCE: APPA, 2000.

infrastructure capacity, and learning and growth objectives were to enhance information management and close skills gaps (Kaplan and Norton, 2001).

The Association of Higher Education Facilities Officers (APPA) developed a Strategic Assessment Model for facilities management that incorporates the four perspectives of the BSC and identifies quantitative performance indicators and qualitative criteria for evaluating the performance of a facilities management organization in each of the scorecard perspectives. The purpose of the Strategic Assessment Model is to help facilities professionals assess their organizations and carry out a continuous improvement program (APPA, 2000). The significance of the model is that many of the measures it incorporates relate to a facilities portfolio, not individual structures. Sample performance indicators used in the Strategic Assessment Model are shown in Table 4.1; other indicators may be more useful for other organizations.

The measures developed to evaluate the performance of the organization, operating units, or individuals in meeting strategic objectives must be developed internally. Heerwagen (2002) provides a set of criteria that organizations of any kind can use in selecting appropriate and useful performance measures:

- *Value.* The measure addresses an important outcome or process and is related to the mission, goals, and objectives of the organization and operating unit.
- *Reliability.* Repeated efforts to measure a phenomenon reach the same results.
- *Validity.* The measure is a good indicator of the outcome of interest (it measures what it purports to measure).

- *Logical connection.* The outcome of interest can be connected logically, or from existing research data, to the program.
- *Ease of gathering data.* The data should be obtainable with minimal extra cost or effort.
- *Efficiency.* The overall measurement plan uses the minimal set of measures needed to do the job and enables conclusions to be drawn from the entire data set.
- *Discriminating.* The measures will allow small changes to be identified.

The inclusion of both qualitative and quantitative measures is an essential aspect of an effective performance measurement system. Quantitative measures, such as operating costs per square foot of a facility, are often readily available and reproducible. Many people regard quantitative data as hard evidence and qualitative data as soft. This distinction is often interpreted to mean that quantitative data are better, when, in fact, they are just different.

The operating environment within which performance measures are applied also impacts the types of measures developed and their utility. Private-sector organizations typically have control over their processes, people, financial resources, and the allocation of resources to meet their internally established objectives. They can design their budgeting and accounting systems to support the types of data needed to evaluate performance in relation to organizational objectives and achievement of mission. Governmental organizations operate under the constraints discussed in Chapter 1.

EVALUATIONS AND CONTINUOUS FEEDBACK

Performance measures are of limited value unless they are used in conjunction with formal and continuous feedback, or evaluation, processes. Evaluations have been defined as the systematic assessment of the operation and/or the outcomes of a program or policy, compared with a set of explicit or implicit standards, as a means of contributing to the improvement of the program or policy (Weiss, 1998). Evaluations of people and processes can help determine if the organization's mission is being achieved and if strategic objectives are being met. They can also serve other purposes important to decision makers and managers:

- *Midcourse corrections.* Evaluations can be used to provide feedback early in program implementation to identify what is happening so that changes can be made before problems become serious and less amenable to correction.
- *Deciding whether to keep, abandon, or change a program.* Evaluations provide data on what the program has accomplished to date and whether or not these accomplishments are in line with goals. If not, decisions can

be made about program improvements to increase effectiveness, or a decision may be made to reduce or abandon the program altogether.
- *Testing a new program idea.* In many instances, new programs start as demonstration projects or experiments. Early evaluation can be used to identify which aspects of the program are succeeding and which aspects may need further development before the program is implemented at a larger scale.
- *Choosing among alternatives.* In cases where several different methods or programs are being tried out, evaluation can provide substantive feedback on which alternative is achieving the best combination of results overall (Weiss, 1998; Heerwagen, 2002).

Continuous evaluation and feedback on processes and investments are essential to controlling and improving them. Feedback can be positive or negative, take many forms, and be used over various timescales. It can be used to bridge the relatively static nature of facilities and the dynamic nature of facilities requirements. Short-term feedback is widely used by organizations of all types to answer questions such as, Was the project completed within budget? Was it operational on schedule? Does it work? If not, why not? Techniques for receiving short-term feedback include real-time assessment, systems optimization, value engineering alternatives, construction realization, and postoccupancy evaluations.

Because of the long-term nature of facilities themselves, longer-term feedback also is needed to identify methods to reduce facility transaction and operating costs and to improve decision criteria and processes. Did the facility investment meet the organizational objectives? Did it correct an operational problem? Reduce long-term operating and maintenance costs? Contribute to a more flexible portfolio? Satisfy users?

In best-practice organizations, the performance of projects, processes, people, business units, physical assets, investments, and the organization as a whole are continuously monitored and evaluated over both the short and long term using performance measures and a variety of feedback processes. An internal set of performance measures designed to capture good performance at levels that directly or indirectly contribute to the desired objectives is developed. Specific elements of performance for individuals, operating units, and contractors can thereafter be tracked and evaluated, providing a strong accountability process in the short and longer term.

FORMS OF FEEDBACK

Project Management

Weekly, monthly, and quarterly control reports for tracking the progress of a project during planning, design, and construction are standard operating procedure in most organizations. The performances of the project manager, operating units, and contractors are also monitored throughout the process. In some cases, the senior person responsible for the project may be required to meet with the board of directors to demonstrate that the project is proceeding according to plan. Adjustments in schedule and personnel may be made weekly or even daily as it becomes apparent that changes are needed. Once a facility project is completed, a more extensive evaluation is made of what worked well and where improvements in processes and project teams are needed for future projects.

Internal Audits

At the Public Service Company of New Mexico, feedback on major facility projects is provided by the project managers and by an internal operating group of auditors with a direct line to the board of directors. The innovative aspect of this approach is that the internal auditors function as internal management consultants rather than as a policing function. The auditing group is tasked with providing management efficiency studies as well as monitoring active projects. In some cases, other operating groups voluntarily call in the audit group to review processes and activities.

The auditing group is staffed by a core group of individuals with accounting and auditing backgrounds, as well as operational backgrounds. Rotational assignments are available for people with operational backgrounds, allowing them to return to their operational function with an audit discipline and some on-the-job training in financial concepts and business skills.

360-Degree Review and Feedback

The Johnson and Johnson Company reported using a process in which all of the participants in a specific project—in-house staff, the architectural and engineering firm, and the construction management firm—rate each other's performance. The project sponsor is also involved in rating the performance of the project. A formalized survey with a 1-5 rating scale for safety, quality, schedule, and other factors is used. All of the information is compiled in a lessons-learned database that can be used to improve processes and performance in subsequent projects.

Peer Review

Independent peer review is often used by owners and project managers of complex engineering projects as a means of quality assurance. It is a method of providing an outside, independent perspective—employing qualified professionals with related experience—to identify issues that may have been missed and to reevaluate decisions to assure that the best alternative has been chosen (NRC,

2003a). Peer review can also be used to introduce more innovation into a project and to overcome traditional impediments to improvement.

Peer review processes typically result in changes in plans: Participants identify better ways of operating as a result of listening to their peers. They can also be useful in bringing together private- and public-sector representatives to interact in a constructive way; in educating junior and mid-level people by exposing them to knowledgeable, more senior people; and in providing a forum for deriving and discussing improved performance measures.

Peer review efforts have been directed at the expensive and urgent challenges of reconstruction at six airports: Buffalo, Boston, New York LaGuardia, Syracuse, Washington Dulles, and Washington National. The peer reviews focused on idea generation and information exchange: Agendas were generated by a discussion among the peers, not presented ahead of time for review, amendment, and ratification. The main results were changes in plans as the airports found better ways of operating when they exchanged information. Cost savings also resulted but were not the primary objective in this instance.

Postoccupancy Evaluation

Postoccupancy evaluation (POE) is based on the idea that better living and work space can be designed by asking users how well the facility they are occupying satisfies their needs. POE is a process for systematically evaluating the performance of buildings after they have been built and occupied for some time. It focuses on the requirements of building occupants, including health, safety, security, functionality, efficiency, comfort, aesthetic quality, and satisfaction (FFC, 2001b).

POE efforts in the United States and abroad have focused on government and other public buildings from the 1960s until today. Some private-sector organizations in the United States began instituting POEs following publication of a 1985 report by Michael Brill et al., *Using Office Design to Increase Productivity*. A number of organizations have since used POE as a tool for improving, innovating, or otherwise initiating strategic workspace changes (FFC, 2001b).

Longer-Term Feedback

The long-term nature of facilities investments, the longevity of facilities themselves, and the continuously changing operations of organizations require long-term as well as short-term evaluations and feedback. Does the facility investment meet organizational objectives? Correct an operational problem? Reduce long-term operating and maintenance costs? Contribute to change management? These and other questions can only be answered through long-term feedback, both continuous and periodic.

Long-term feedback requires a cradle-to-grave monitoring and evaluation system supported by integrated databases and formal feed-forward mechanisms. Such systems, sometimes referred to as lessons-learned programs, are designed to collect, archive, and share information about successes and failures in processes, products, and other building-related areas for the purpose of improving the quality and life-cycle cost of future buildings (FFC, 2001b).

The Disney Corporation provides one model from the private sector: It has been evaluating everything it does since the 1970s. Disney has at least three evaluation programs and three corresponding databases. One program evaluates the performance of materials and equipment, and the findings are recorded in a technical database. A second program focuses on predictors of customers' intentions to return, Disney's key business driver. A third is aimed at refining programming guidelines and rules of thumb. The databases are not formally linked but are used extensively during design and renovation projects. By using these databases, the design and engineering team can improve future facilities based on past experience and research. For example, streets can be designed to allow the optimal number of visitors to prevent overcrowding and stimulate gift shopping, key factors in Disney's future success (FFC, 2001b).

In best-practice organizations, once a facility is in operation, evaluation and feedback are employed to track operating costs and other factors over the longer term to determine if investment and organizational objectives are being met. They also measure and evaluate the performance of their entire portfolio of facilities in relation to organizational objectives. Such evaluations and feedback may often drive changes in the decision-making process itself.

ACCOUNTABILITY

Accountability has been defined as the relationship between those who control or manage an entity and those who possess formal power over them. It requires the accountable party to provide an explanation or a satisfactory reason for his or her activities and the results of efforts to achieve the specified tasks or objectives. The process of accountability includes rendering an account of or explaining one's actions to those in authority or formal power so that they may assess performance, make a judgment, and take action (GASB, 1994).

Private-sector organizations operate on a risk/reward basis, rewarding those individuals or operating units accountable for successful execution of an idea, project, or other activity and penalizing those accountable for less than successful efforts within their control. They link accountability, responsibility, and authority when making and implementing facility investment decisions. If an individual or operating group is to be held accountable for a decision to proceed or not to proceed with a facility investment, or for the execution of a project, that individual or operating group is given the appropriate level of authority and resources to meet their responsibilities. At the same time, they are held accountable for the results, whether positive or negative, and rewarded or punished accordingly.

Typically, at higher levels of management the responsibilities and authorities of individuals increase and they become more accountable for the performance of their operating unit or the entire organization. In the case of a facility investment proposal, the reviewing body that decides to proceed or not to proceed may be held accountable for the results of that decision in relation to the achievement of operational or organizational objectives.

Establishing accountability in the federal government, where decision-making authority and responsibility are spread throughout the executive and legislative branches, is a complex proposition that is discussed more fully in Chapter 6.

INCENTIVES

In private-sector organizations, incentives are created to motivate in-house individuals and operating units to meet organizational objectives. The incentives are often, but not always, financial. For instance, for groups of individuals, bonuses may be linked to how their operating unit contributes to meeting organizational objectives. Bonuses for senior executives may be tied to overall corporate performance, while bonuses for operations staff may be tied to project performance. Incentives are also created in less significant but important ways: through recognition and awarding additional vacation days, priority parking spaces, plaques and trophies, and the like.

In one firm interviewed, every partner owns a percentage of all facility or real estate projects and thus has an incentive for monitoring the performance of the entire portfolio of projects. Quarterly reviews are held with investors. At regular intervals, the oversight committee reviews the performance of individual projects. The intervals are shorter if performance or market conditions deteriorate; if a project is in serious trouble, the monitoring is constant.

In the public sector, creating incentives based on financial reward is a more difficult prospect and raises concerns about how such incentives can be appropriately designed. A wide range of nonfinancial incentives, including recognition, can also be used. These issues are more fully addressed in Chapter 6.

PRINCIPLES AND POLICIES FROM BEST-PRACTICE ORGANIZATIONS

Based on a consolidation of research, interviews, briefings, and the committee members' individual and collective experience, the committee found that best-practice organizations that successfully manage facilities investments operate under a number of principles and policies when they make decisions (all 10 principles/policies are repeated in Chapter 6):

Principle/Policy. Best-practice organizations establish a framework of procedures, required information, and valuation criteria that creates an

effective decision-making environment and that provides a basis for measuring and improving the outcomes of facilities investments. The components of the framework are understood and used by all leadership and management levels.[5]

In best-practice organizations, effective decision making for facilities investments is related to managing a free exchange of information among the various stakeholders, particularly those who might be skeptical about a proposed investment. Open communications ensure that those who need to know and who can best critique a proposal have access at a sufficiently early stage to provide information and insights that can be constructively used to produce a better proposal. The more open the process, the more likely it is that errors in fact or methodology will be uncovered. Such organizations use additional framework components—performance measures, feedback processes, accountability, and incentives—to measure, adjust, and improve decision-making processes, management practices, and the results or outcomes of decisions.

Principle/Policy. Best-practice organizations use performance measures in conjunction with both periodic and continuous, long-term feedback to evaluate the results of facilities investments and to improve the decision-making process itself.

The notion of continuous process monitoring and feedback is built on the recognition that however effectively one plans, unforeseen events, unintended consequences, and change will occur. Best-practice organizations establish baselines and performance measures to monitor processes and the results, or outcomes, of those processes in relation to organizational objectives. Both quantitative and qualitative measures are used, and the measures are tailored to the organization's culture, mission, and objectives.

Continuous feedback on processes and investments can be positive or negative, can take many forms, and can be used over various timescales. Short-term feedback is widely used by organizations of all types. Not as widely used is longer-term feedback, which is useful in identifying methods to reduce facility transaction and operating costs and for improving decision criteria and processes.

Principle/Policy. Best-practice organizations link accountability, responsibility, and authority when making and implementing facilities investment decisions.

Private-sector organizations operate on a risk/reward basis, rewarding those individuals or operating units accountable for successful execution of an idea, project, or other activity, and penalizing those accountable for less than success-

---

[5]This principle/policy and that in the Executive Summary and at the end of Chapter 3 together form Principle/Policy 1 in Chapter 6.

ful efforts within their control. If an individual or operating group is to be held accountable for a specific result, it is given the appropriate level of authority and resources to meet its responsibilities.

Principle/Policy. Best-practice organizations motivate employees as individuals and as groups to meet or exceed accepted levels of performance by establishing incentives that encourage effective decision making and reward extraordinary performance.

In best-practice organizations, incentives are created to motivate individuals and operating units to meet organizational objectives. The incentives are primarily financial but also include recognition and other less significant but meaningful rewards for superior performance.

5

Alternative Approaches for Acquiring Federal Facilities

BACKGROUND

Facilities investments typically require substantial up-front funding for design, construction, or outright purchase. The benefits from such investments may not begin to accrue for 2 or more years as facilities are constructed or renovated.

Private-sector organizations typically finance facilities investments by borrowing all or a portion of the required funds from a bank or other lending institution, by using their own financial resources, or by using some form of third-party financing or equity arrangement. Additional arrangements are routinely used, such as alliances with other firms, joint ventures, sale-and-leaseback, and public-private partnerships. All involve varying levels of risk, and some incur debt. A loan or other commitment is typically repaid over time, allowing the organization to receive value from the investment before the debt is fully repaid.

In the federal government, significant facilities investments are primarily funded from the annual budget. Individual departments and agencies may not borrow funds or otherwise incur debt to finance facilities.[1] They must receive authorization from Congress for funding to cover the full, up-front (design and construction or purchase) costs in a specific fiscal year budget.

Although the annual budget is the primary source of funding, a number of alternative approaches for acquiring facilities are being used by federal departments and agencies, on a case-by-case basis under special agency-specific legis-

---

[1]The Tennessee Valley Authority, an independent, wholly owned corporation (not a department or agency) of the federal government, has the authority to issue bonds and notes and thus to incur debt and other financial obligations (GAO, 2003h).

lation. This chapter first discusses issues related to full, up-front funding of facilities, including procedures for budget scorekeeping, and issues related to alternative acquisition approaches. A number of alternative approaches, including public-private partnerships, capital acquisition funds, trust funds, sale-and-leaseback arrangements, outleasing, real property exchanges, and shared facilities, are then described and analyzed in greater detail. The chapter concludes with a summary and a recommendation for the federal government.

ISSUES RELATED TO FULL UP-FRONT FUNDING OF FACILITIES

The requirement for full, up-front funding of federal facilities is intended to (1) give adequate scrutiny to the initial costs and proposed benefits of an investment; (2) avoid the risk of allowing projects to be started through incremental funding before they are adequately scrutinized; (3) give Congress the flexibility to respond to changing circumstances and priorities; (4) provide for transparency in the budget by making sure the investment proposal is understandable to a range of constituencies, and (5) allow for the informed participation of those constituencies.

Under current procedures, the budget authority associated with requests to design and construct a new facility, to fund the major renovation of an existing facility, or to purchase a facility outright are scored up front in the year requested even though the actual costs may be incurred over several years. Thus, the projected costs, which may easily run more than $50 million per facility, are counted against the agency's overall budget request for a given fiscal year.

The requirement for full, up-front funding, however, typically results in a spike in a department's or agency's budget request. If it is to stay within its spending cap, a request for a significant facility investment will force cuts in other programs or activities within the department or agency, causing tension among the various in-house decision-making and operating groups.

The focus on first costs of facilities investments is reinforced by the budget scorekeeping rules mandated as part of the Budget Enforcement Act of 1990. "Scorekeeping" is a process for estimating the budgetary effects of pending and enacted legislation and comparing those effects with limits set in the budget resolution or legislation. It has no analogue in the private sector.

Scoring facilities costs up front is intended to provide the transparency needed for effective congressional and public oversight. The objectives are to (1) highlight the full costs of decisions when they are being made; (2) discourage the undertaking of investments that are not cost-effective; (3) protect congressional control over federal spending; (4) see how legislation fits into the overall plan for federal spending; and (5) determine if any ceilings in those plans have been breached (CBO, 2003).

In actuality, up-front scoring of major facility proposals does not disclose the full costs of facility investment decisions; only the projected design and construction costs of facilities are transparent. Facilities operation, maintenance, repair, and disposal costs are accounted for in different functional areas of the budget and are not identifiable for specific facilities.

Scorekeeping procedures also create incentives for agencies to drive down the first costs of facilities investments—even if it may increase the life-cycle costs—in order to lessen their apparent impact on the current year budget. In rewarding such behavior, the scorekeeping procedures can indirectly increase the long-term operation and maintenance costs of facilities—that is, 90 to 95 percent of their life-cycle costs—and decrease the staffing efficiencies that might result from additional initial investment.

Another consequence of the scorekeeping procedures for major proposals is that

> [a]gencies faced with the upfront costs of acquiring new capital assets—facilities and equipment—often have the option of continuing to produce goods and services using what they have even if it is old and obsolete. Although the approach can increase the costs of producing output in the long run, it holds down budgetary costs in the short term. (CBO, 2003, p. 21)

ISSUES RELATED TO THE USE OF ALTERNATIVE APPROACHES FOR ACQUIRING FACILITIES

Recognizing some of the difficulties of providing adequate funding for required facilities investments through the annual budget process, legislation has been enacted over the years on a case-by-case basis for individual departments and agencies to allow the use of alternative approaches for acquiring facilities. Legislation allowing the use of these approaches on a government-wide basis has not occurred for a variety of reasons. First, the use of alternative approaches for acquiring federal facilities creates a tension between government-wide oversight groups—Congress, the OMB, the CBO—and the line agencies. As noted by the GAO,

> From an agency's perspective, meeting capital needs through alternative funding approaches (i.e., not full funding) can be very attractive because the agency can obtain the capital asset without first having to secure sufficient appropriations to cover the full cost of the asset. Depending on the financing approach, an agency may spread the asset cost over a number of years or may never even incur a monetary cost that is recognized in the budget. From a government wide perspective however . . . the costs associated with these financing approaches may be greater than with full up-front budget authority (GAO, 2003a, p. 1).

The use of alternative approaches for facilities investments raises a second issue: Who retains the proceeds from the sale or leasing of properties, called "offsetting collections," for federal budget purposes? Under federal procedures, any proceeds realized by the sale of federal buildings or properties are returned to the general treasury unless special legislation has been enacted.

For individual federal agencies, authorization to retain and use the proceeds for the sale or leasing of property could provide incentives to find more cost-effective ways to manage their facilities portfolios. On the other hand, from a government-wide perspective, retention of proceeds could spur some agencies to sell properties that are required for long-term mission support to generate funding for more short-term needs.

A third issue relates to the public nature of facilities investments. Because federal facilities are located in all states and most communities, the perspectives of state and local governments and constituencies must be accounted for when alternative funding approaches are considered. Federal departments and agencies are not typically subject to local zoning and land use controls when siting facilities. What may appear to be cost-effective from a departmental or agency perspective may significantly affect the surrounding community and may not appear to be cost-effective or desirable from the perspective of the locality and citizenry.

In some cases, the short-term benefits of some alternative approaches may not outweigh the long-term, quantifiable costs. Still, in certain circumstances, such approaches can provide other, less tangible benefits to the public. These include the preservation and upkeep of historic properties, investment and occupancy of buildings in downtown and inner city neighborhoods, and more convenient access to services.

A number of alternative approaches currently in use by federal agencies are described below. Each approach has advantages and disadvantages for particular types of organizations and types of facilities. None can guarantee effective management absent agreed-upon performance measures, feedback procedures, and well-trained staff. It should be noted that state and local governments have also developed innovative funding approaches that could be adapted to the federal government. Identification and evaluation of such approaches were beyond the scope of this particular report, but such an analysis by the appropriate federal entities is warranted.

Public-Private Partnerships

In general, the concept underlying public-private partnerships is to utilize the untapped value of real property. One type of public-private partnership is a project for which the private sector provides cash and financing ability to renovate or redevelop real property contributed by the federal government. Once such a project is completed, both partners share in the net cash flow that is generated (PriceWaterhouse, 1993). A more general conception of a public-private partnership includes sharing of other responsibilities such as project planning and initiation, design, construction, and operations management. The extent of the allocation between the public and private sectors in any of these areas depends on the specific agreement between the two sectors. Two examples of current programs follow.

Department of Veterans Affairs Enhanced-Use Leasing Authority

In 1991, Congress gave the Department of Veterans Affairs (VA) authority to lease underused property and facilities to private or other public entities for up to 35 years in return for cash or in-kind consideration, such as services, goods, equipment, or facilities. The basic intent of this authority was to increase the agency's flexibility to utilize underused assets that could not or would not otherwise be disposed of. In 1999 this authority was extended for 10 years, and the applicable lease term was extended to 75 years. As part of this extension, the VA was also allowed to lease properties for the sole purpose of generating revenues to provide better services for veterans, and the agency was also allowed to make capital contributions to joint ventures on agency properties (FFC, 2001a).

The basic process for using the enhanced authority requires local VA offices to develop a business plan. The organization that initiates a successful proposal can retain the net proceeds from an enhanced-use lease agreement. Public hearings are held on any proposal, after which the proposal moves to the VA headquarters for review. Projects valued at more than $4 million must be approved by the OMB and go through the *Federal Register* process before a Request for Proposal is issued. Once a proposal is negotiated, Congress is notified and the VA must wait 30 days before entering into a lease.

As of 2003, the authority has been used in more than 27 agreements (GAO, 2003a). As an example, in Texas a private developer constructed a regional VA office on the VA's medical campus, and the agency in turn leased land on the campus to the private developer so that it could construct commercial buildings with space rented out to private businesses (FFC, 2001a). Enhanced-use leasing authority recently has been granted to the National Aeronautics and Space Administration and the Department of Defense, although the specific procedures and requirements of their authorities vary from those of the VA.

National Park Service Concessions Program

For many years the Department of the Interior's National Park Service has entered into long-term agreements with private entities to manage certain facilities on government-owned properties. In 1998, 630 concessionaires provided services grossing $765 million in revenues. Almost two-thirds of this total came from the 73 concessionaires that provided lodging. A change in the law that same year increased competition for those concessions and created an advisory board to the Secretary of the Interior to suggest ways for improving the process.

Potential benefits of public-private partnerships include the conversion of properties that might currently drain public resources into useful and productive facilities that provide net cash inflow to federal agencies. Partnerships also can attract private funding sources for renovations and repairs. In addition, the introduction of private-sector, profit-motivated entities may increase the efficiency with which existing properties are managed (GAO, 2001d).

The up-front and long-term costs of public-private partnerships vary. Considerations for the government include the contract agreements for occupancy by federal entities, restrictions on the use of the property, liability for the actions of any particular lessee, and the leasehold interests of the government in relation to any lender of the nongovernmental partner. Any federal public-private partnership is subject to budget scorekeeping rules. There could be cases where a financial analysis of a transaction by an agency is at variance with the scorekeeping analysis as determined by the OMB or the CBO—that is, the agency analysis might show a net benefit, while the scorekeeping analysis might show a net cost.[2]

Although the study committee supports more widespread use of public-private partnerships, it offers some caveats. The public interest must be considered before entering any partnership. Even if a transaction is viable from the private perspective, there should be sufficient financial returns to the government to warrant it. Public objectives related to accessibility, the environment, and historic preservation should not be compromised. Strict controls to avoid conflict of interest and other forms of potential or actual corruption are required. All of these factors should be weighed in a partnership feasibility analysis because they may argue against a partnership that looks attractive on more narrow financial grounds.

Program design must also take these factors into account. The VA program, for example, has many checkpoints, including public hearings at the local level, that must be passed through before an enhanced-use lease agreement becomes final. No matter how well designed an agreement may be, poor implementation and execution can undo the benefits or, worse, lead to losses. The Park Service's concessions program, for example, has in the past suffered from the fact that agency staff overseeing the contracts are often inadequately trained and have lacked basic business analysis and management skills (GAO, 2000a). Lack of performance-based contracts in that agency is another problem, as is a muddling of lines of authority and accountability from project-level staff to higher-level agency executives. For example, the chief of concessions has no direct authority over staff in the individual park units who manage individual concessions. Although these are particular examples relating to the Park Service, they illustrate the kinds of issues that must generally be resolved satisfactorily in any broadening of the use of public-private partnerships in the federal government.

Capital Acquisition Funds

Proposed in the Report of the President's Commission to Study Capital Budgeting (1999) and in the President's Budget for FY 2004, capital acquisition funds (CAFs) are accounts that would receive appropriations for large capital projects,

---

[2] See the CBO report *Budgetary Treatment of Leases and Public/Private Ventures* (2003) for a full discussion of these points.

appropriations that are now made to individual operating units within federal agencies. The CAFs would use the authority represented by those appropriations to borrow against the general fund and would acquire the assets on behalf of the operating units within the agencies, charging those units rent on the facilities equal to the cost of debt service on the relevant project. Thus, if an agency wanted a new capital project, Congress could allow the agency to borrow money through its CAF to purchase it. Agency programs would then repay the fund, based on their use each year.

Although proposed as agency-wide, a CAF could be applied at a higher level across agencies; for example, appropriations committees could appropriate to a CAF established for all agencies under the purview of their particular committee. A CAF would not replace the General Services Administration (GSA) revolving fund (the Federal Building Fund[3]). Instead, agencies would use their CAF only if their office space acquisition could be done more effectively and efficiently than through GSA.

CAFs have not yet been used in the federal government, and how they would operate is still unclear. It would seem that oversight and management of such funds should reside in a central budget organization such as OMB. Under the proposal in the President's Budget for FY 2004, departments would no longer receive separate appropriations for support services and capital assets but would create a fund at each department that program managers would use to buy facilities-related requirements. Managers could then buy support services from the government or the private sector with the funds. Although the proposal is broader than a CAF alone because it covers noncapital services in addition to capital programs, CAFs are explicitly a part of the proposal.

A CAF has several perceived advantages over current agency methods of capital funding. First, it would require capital asset coordination and planning across agency operating units. Second, a CAF could smooth out funding and expenditure spikes that occur when individual units have especially large periodic capital requests. Finally, because operating units would be charged annual "rent" (representing debt service and other asset overhead), a CAF could lead to more accurate allocation of asset costs to affected parties within agencies, giving asset managers incentives to make more efficient decisions.

The existence of a CAF by itself would not ensure good implementation and management. As proposed, a CAF is an additional layer of administration that could complicate program management rather than streamline it. Issues to be worked out include the relationship between a CAF and the GSA Federal Building Fund, the managerial relationship between a CAF and individual operating units within agencies, and the status and treatment of CAF activities within the current overall operating budget.

---

[3]The Federal Building Fund was established under the Federal Property and Administrative Services Act of 1949.

The measure could also present challenges for agencies that own extensive property. Whereas currently a congressional appropriation for a capital project is simply added to the budget of the fiscal year in which it is appropriated, under a CAF, as outlined in the new proposal, an agency's capital acquisition fund would borrow the needed money, and that money would be gradually paid off by the agency programs that used it. Assigning costs in this way would make projects appear more expensive. That is an intended consequence meant to ensure the overhead costs of a capital project are more explicit and borne by the managers and users of that project. Indeed, according to the CBO, such an approach works on the premise that the federal budget should recognize capital costs up front when the decision to invest is made while spreading those costs out over time in program managers' budgets (CBO, 2003).

Despite these caveats and issues, the committee believes that CAFs offer an opportunity to fulfill facilities-related requirements more cost effectively and efficiently. The committee supports implementation of pilot programs using CAFs to determine if their promise can be realized in the federal operating environment.

Dedicated Funding, Trust Funds, and Earmarked Receipts

Dedicated funding refers to any mechanism whereby resources are committed to a specific purpose in advance of any actual spending or activity and which in some way guarantees that those resources will actually be spent according to that initial commitment. A variety of mechanisms are used to ensure dedicated funding. A simple one is a direct mandate, perhaps contained in the charter of an organization that contractually or legally forces an entity to spend certain monies in a specified way.

More widely used are the devices of trust funds and earmarked receipts. In the private sector a person creates a trust fund using his or her assets to benefit specific individuals. The creator of the trust names a trustee who has fiduciary responsibility for managing the designated assets in accord with the stipulations of the trust (GAO, 2001a). In the federal government, Congress creates a federal trust fund in law and designates a funding source to benefit specified groups or individuals or, in some cases, itself. The Treasury Department and the OMB determine the budgetary designation as a trust fund when a law both earmarks receipts and identifies the account as a trust fund account (GAO, 2001a).

Earmarked receipts are collections that are stipulated by law as being dedicated to a specific fund or purpose. Earmarked funds do not always go to trust funds. They also are deposited into entities such as public enterprise funds, which often have the same purposes as trust funds but are not designated as such. Two examples of earmarked funds are the Nuclear Waste Fund and the Postal Service Fund. Examples of federal trust funds include Social Security, Medicare, and the Highway Trust Fund.

There are two types of federal trust funds: (1) revolving funds, which support a cycle of businesslike operations in which earmarked receipts are derived mainly from revenues generated by those businesslike activities, making the relationship between the sources and uses of funds relatively clear, and (2) nonrevolving funds, in which the earmarked receipts are not generated by businesslike activities but come from periodic revenues such as annual appropriations and a variety of other sources, from cigarette and payroll taxes to customs duties (GAO, 2001a).

Designation as a trust fund does not impose a greater commitment on the part of the government to carry out that activity than it has to carry out other activities. Although special constituencies may create pressure to spend earmarked revenues, the federal government does not have fiduciary responsibility to the trust beneficiaries in establishing and operating a trust fund, revolving or otherwise. While the law establishing a given trust fund does govern the collection and disbursement of revenues going into that fund, Congress can change the law to change the terms of how much money is collected, how much is disbursed, to whom it is disbursed, or the purposes for which the funds are used. In addition, in most cases the federal government has custody and control of the funds and the earnings on those funds.

One example of a trust fund used for facilities acquisition and investment is the U.S. Mint Public Enterprise Fund. Established in 1996, this fund allows all receipts from the Mint's operations to be deposited into an account from which all operations are then funded. Such operations include "the acquisition or replacement of equipment, the renovation or modernization of facilities, and the construction or acquisition of new buildings" (P.L. 104-52). The fund is unique in that the Mint's operations are exempted from the Federal Acquisition Regulations, which cover government procurements and public contracts. The exemption allows the Mint to operate more like a private-sector entity, thus gaining the flexibility and efficiency that purportedly accrue to such entities. Under this exemption, the Mint itself determined that it also had statutory lease authority and thus did not fall under the leasing rules set forth by the General Services Administration (OIG, 2002).

Trust funds, earmarked funds, and charter mandates have the advantage of being relatively simple in concept and focused on a single aim: provision of dedicated and sufficient funds for an intended purpose. In that sense they have performed well. The public readily understands the concept, and when one of these mechanisms exists, it tends to create momentum toward keeping funding at least at a certain minimum level. Similar results can accrue to public enterprise funds and other special funds in the federal budget receiving earmarked funds.

However, the existence of a trust fund or other mandate does not guarantee that funds or facilities will be well managed. An investigation of the U.S. Mint found that the agency was leasing too much space for its needs, was not following prudent business practices in its leasing arrangements, and in general had weak management controls. Thus, having a dedicated trust fund, which in this case gave extra operational flexibility, is not a replacement for good management (OIG, 2002).

There are also issues surrounding the interpretation of trust fund balance information. A fund may be running a surplus, something that may be interpreted as indicating a healthy program. Yet the program may not be financially or managerially sustainable in actual fact if the trust fund flows are not designed with long-range needs in mind and if program funds are not soundly administered. Similarly, a deficit does not necessarily indicate a troubled program. Even if it does, the response may be to simply add more funds without addressing fundamental problems.

Trust funds, earmarked funds, and special funds are widely used in the federal government. In FY 1999 half of federal receipts went into trust funds, and 130 of them existed at that time—120 nonrevolving and 10 revolving. Issues related to their continued use include whether they should be renamed to avoid confusion on the part of the public with private sector trust funds, whether there should or could be some tightening of terms to make them more like private trust funds, whether information provided on them should be revamped to reveal more about program operations than a mere fund balance, the strength of the link between the source of the funds and their use, and how the use of funds is linked to underlying program management regimes—that is, the transparency of the funding (GAO, 2001a).

Sale-and-Leaseback Arrangements

Sale-and-leaseback arrangements are routinely used in the private sector. The owner of a building sells it to another company or entity and then leases it back for a specified time period. At the end of that time, the original owner buys the building back. This type of arrangement allows the original owner to raise capital and still retain use of the building, in essence temporarily borrowing funds that can then be used for other purposes (Groppelli and Nikbakht, 2000).

A sale-and-leaseback arrangement offers few, if any, incentives for a federal agency unless it can retain the sale proceeds and use them to achieve some benefit or purpose that is not being funded through its annual budget. In at least one instance, the GSA was granted authority to retain the proceeds if it entered into a sale-and-leaseback arrangement. In this case, the GSA had planned to excess a Class C office building in West Virginia after the federal tenants in the building moved to a new courthouse. In the interim, the Social Security Administration contacted the GSA about moving into the Class C space in order to better serve the public by consolidating its functions with those of the West Virginia Disability Determination Agency. Legislation was enacted allowing GSA to sell the building and retain the proceeds for the Federal Buildings Fund. The new owner committed $11 million to upgrade the building to Class A office space; in turn,

GSA committed to leasing a portion of the building back for 20 years, thus assuring the owner of a stream of revenue to pay back its investment. Both the GSA and the Social Security Administration claimed immediate benefits from this arrangement. However, the GAO expressed concern whether this arrangement would be cost-effective in the long term (GAO, 2003a).

Outleasing

Under an outleasing arrangement, a federal agency leases all or a portion of a facility to a private-sector or not-for-profit organization. The lessee assumes the maintenance and repair costs of that space and in some cases invests in renovations. In essence, the federal agency becomes a landlord to nonfederal entities.

Outleasing arrangements have been used by the GSA, the Coast Guard, and possibly other agencies, for some underutilized and historic properties (GAO, 2003a). The Coast Guard, for instance, has outleased and divested 28 historic lighthouses in the State of Maine to organizations that will ensure the lighthouses are repaired and maintained. Under this arrangement, the Coast Guard receives some income from the lighthouses that can be used to offset expenses at other historic properties and avoids annual maintenance and repair costs of $3 to $5 million. It thus receives a benefit from properties that are no longer integral to its mission. The public benefits in that the properties are preserved for posterity and in a better state of repair. The GSA has used outleasing to gain similar types of benefits from other historic properties, such as customhouses (GAO, 2003a).

Clearly such arrangements raise questions about the relative costs and benefits of selling excess historic and underutilized facilities outright and maintaining some control and stewardship over heritage properties. They also raise issues related to local land use control and interests. The costs and benefits, financial and intangible, will vary case by case but offer the potential for improved stewardship of federal properties.

Real Property Exchanges

Sometimes land and buildings owned by a federal agency have a greater value to another entity than to the agency itself. On occasion, the GSA, the Air Force, other military services, and possibly other agencies, have been able to exchange real property with a private developer or a state or local government in return for a different piece of property or facilities. Such exchanges are different from public-private partnerships in that they typically do not involve an exchange of funds or competitive bidding; there are a limited number of potential special-purpose exchanges; congressional oversight is more limited; and such exchanges are not reflected in the federal budget (GAO, 2003a).

In one instance, the Army Reserves conveyed approximately 11 acres of land used for training activities to a private-sector developer that required the land to build a road for access to a new development. In exchange, the developer constructed a new fire station for the Army Reserves, to replace an older, less modern one (GAO, 2003a).

In another example, the GSA conveyed two small parking areas and a partially vacant, deteriorating historic property to the city of Albuquerque, New Mexico, in exchange for a large parking garage proximate to other federal buildings (GAO, 2003a).

In a third instance, legislation was enacted that authorized the Air Force to convey land it owned on the Los Angeles Air Force Base to a private developer in exchange for the design and construction of a new 560,000-square-foot space and missile systems center on the base. The new center replaced two outdated buildings that were vulnerable to earthquakes.

In these cases and others, the federal agencies involved were able to exchange real property for other land or buildings that provided greater benefit to the agency without having to use funds from their annual budget appropriation. From a government-wide perspective, real property exchanges raise issues about the property valuation procedures used, the fair market value of the property if a competitive bidding process is not used, the sufficiency of congressional oversight, and how to reflect such exchanges in the federal budget.

Shared Facilities

"Shared facilities" refers to the practice of having independent operating entities with large portfolios of facilities share the use and/or management of those facilities in some way. The sharing could apply to information about the facilities, coordination of planning and management, joint oversight, or actual shared use of facilities. As an example, the GSA oversees and coordinates the Government-wide Real Property Information Sharing (GRPIS) program, which allows different federal agencies to share information about facilities under their individual control. Purely voluntary in its participation, GRPIS has so far resulted in the formation of interagency real property councils in several regions of the country; development of an automated inventory of real property; a Web site for sharing information about best practices, ongoing issues, and follow-on initiatives; and joint analyses of common issues in the regions and possible coordinated solutions to those problems (GSA, 1998).

Sharing facilities is a way to extract more utility from a portfolio of facilities. By treating a facility as commonly held rather than individually held, managers can avoid duplication of effort in both current operations and future investments; fully utilize assets that, if used only by the owner of the facility, might be underused; and share costs, making the facility more affordable and manageable.

Disadvantages of facility sharing vary according to what is being shared. The GRPIS program is a voluntary information-sharing program. As the GSA itself notes, while a voluntary program increases the level of trust and perhaps enthusiasm for participation, it also potentially makes for less effective joint action. If more than information is shared and if participation is mandatory, other problems might be introduced. Joint use and management of facilities, for example, can be a costly activity, in terms of both staff time and direct outlays. And depending on how disparate the operating entities are and how diverse the facilities portfolio being managed is, sharing of facilities can make management slower, less responsive, and less effective.

SUMMARY AND A RECOMMENDATION

Based on a consolidation of research, interviews, briefings, and the committee members' collective and individual experience, the committee found that a range of alternative approaches to acquiring federal facilities are used by individual agencies under special legislation specific to the agency. Capital acquisition funds, not yet implemented in any federal agency, offer the potential for improved capital asset coordination and planning across operating units, more accurate cost allocation of assets, and incentives for asset managers to make more cost-effective decisions.

However, each of these approaches has advantages and disadvantages. Successful implementation of alternative approaches requires effective oversight and management by federal employees with the appropriate skills and training.

When implementing an alternative approach, the committee believes that all the potential costs and benefits to federal departments and agencies and the public should be taken into account. The impacts on state and local communities should be accounted for and attempts should be made to balance national, departmental, and agency objectives with those of other public stakeholders.

Taking these caveats into consideration, the committee believes that if alternative approaches for acquiring facilities were carefully applied, their use on a government-wide basis could provide federal departments and agencies with more cost-effective ways to acquire facilities, reinvest in the existing stock, and provide a range of benefits to the public. Pilot programs to test the effectiveness of various approaches and to evaluate their outcomes from national, state, and local perspectives should be implemented as a first step. If changes to the budget scorekeeping rules are required to expand the range of alternative approaches, such changes should be tested through the pilot programs.

Recommendation: In order to leverage available funding, Congress and the administration should encourage and allow more widespread use of alternative approaches for acquiring facilities, such as public-private partnerships and capital acquisition funds.

6

Adapting Principles and Policies from Best-Practice Organizations to the Federal Operating Environment

BACKGROUND

In Chapters 2 through 5, the committee identified principles and policies used by best-practice organizations for facilities investment and management. The committee found these principles and policies to be largely independent of the size and complexity of the organizations, their form (e.g., corporation, partnership), their orientation toward goods or services, and their centralization or decentralization. The practices used to implement the principles and policies, however, vary widely and are tailored to an organization's structure, goals, resources, and culture. In this chapter, the committee addresses how the identified principles and policies from best-practice organizations could be tailored to the structure of the federal government and to the goals, resources, and cultures of its individual departments and agencies.

The chapter first reviews special aspects of the federal operating environment that must be considered in any adaptation of the identified principles and policies. The following section consolidates and reiterates the principles and policies used by best-practice organizations as defined and identified by the committee. Issues and barriers related to adapting these individual precepts for use in the federal operating environment are discussed and one or more recommendations for their adaptation are made. The chapter concludes with the committee's recommended overall strategy for implementation.

SPECIAL ASPECTS OF THE FEDERAL OPERATING ENVIRONMENT

Although many have suggested that the federal government adopt principles, policies, and practices used by private-sector organizations to make the government more "businesslike" in its operations, significant differences in the two environments complicate their direct transfer.

Mission

As long as profits result, a private-sector organization's mission, values, and leadership can remain relatively unchanged for years. In the federal government, the accepted overall goal is to promote the general welfare of the public; federal departments, independent agencies, corporations, and commissions each have multiple missions and programs intended to help achieve the overall goal. However, the electoral process ensures change in executive and legislative leadership on a regular, relatively short-term basis. As the leadership changes, the emphasis placed on meeting particular missions also changes.

The electoral process in the legislative branch and at the top of the executive branch also means that the major participants are acting within a framework of public positions on many of the values and priorities implicit in facilities projects. The time between initial project analysis and decision making and the start of execution can be quite long and span several administrations. Consequently, in the government, accountability for decision making is dispersed among a myriad of stakeholders, some of whom may no longer be with the government by the time decisions for investments are implemented and the facilities are subsequently operated.

The Organizational Structure

In large private-sector organizations, the chain of command between decision makers and operating groups is relatively short, the size of the decision-making group is relatively small, and there are strong commonalities of goals and values among all those involved.

In the federal government the decision-making environment is rather more complex, deriving in part from the separation of powers between the executive and legislative branches and, within the legislative branch, between the Senate and the House of Representatives; the organizing principle of checks and balances at all levels; and the consequently much longer command chains. This system ensures that the many viewpoints, possible outcomes, and consequences of public policy decisions are identified, considered, and accommodated, which can span several administrations.

Rather than operating as a single entity, the federal government operates as a network of distinct but interdependent organizations. Federal facilities investment decisions involve multiple stakeholders, decision makers, and operating groups with differing missions, values, goals, and responsibilities, which may sometimes be overlapping and sometimes conflicting. In this network-like structure, responsibility and authority for decision making are spread throughout the executive and legislative branches and frequently are not directly linked.

Within the federal structure, departments and agencies are somewhat analogous to private-sector organizations. Departments and agencies have specific and varied missions; significant resources at their disposal to achieve those missions; and a variety of decision-making and operating groups—human resources, facilities, research, financial, policy-level and program-level units, public relations, etc.—with differing objectives, responsibilities, and technical knowledge. They have some flexibility in establishing processes for the evaluation of facilities investment proposals, although they must all follow the same procedures for funding requests through the annual budget process.

The answer to the question, What facilities are required? typically begins to be formulated at the department and agency level. It is here that facilities requirements to support organizational operations or meet congressional and presidential directives are identified, alternatives are developed, and analyses of facilities investment proposals are conducted. Trade-offs begin to be made among alternatives for a specific investment proposal, among a range of proposals, and among investments in facilities and other important organizational activities.

Unlike private-sector organizations, federal departments and agencies cannot independently make a final decision to proceed with a significant facility investment or to independently allocate funding for that investment. Instead, they can only recommend that an investment be made and then forward that recommendation to the Office of Management and Budget for its review and to Congress for a final decision. These reviews and approvals involve a set of stakeholders who take a government-wide perspective and whose responsibilities, objectives, and values differ.

The Nature of Federal Facilities Investments

Another distinction between private-sector organizations and the federal government relates to who pays for and who benefits from the facilities and infrastructure in which they invest. Federal facilities investments are funded by the American public and therefore incur costs and confer benefits on a wide spectrum of people and organizations. Such investment decisions must take into account the costs and benefits to the public at large, not just those to a specific agency, department, or organization. The benefits are often qualitative rather than quantitative and can be difficult to measure. The costs and benefits may also differ depending on the level—national, state, regional, local, departmental, or agency.

The wide range of government roles and missions means that each funding proposal for facilities must compete with many more alternatives for public investment, each with quite different measures of social utility. Federal facilities that support public services do not generally operate under easily quantifiable dollar measures of costs, operating margins, and market performance, further complicating simple metrics for making decisions. The commitment of public funds also requires far more transparency in the process than does that of private-sector funds.

Decision-Making Environment

In the federal government, as in many private-sector organizations, requests for funding of particular programs, projects, and initiatives typically exceed available resources. Decision makers in Congress and federal departments and agencies are asked to balance the competing demands of very different programs: Funding for facilities investments must be weighed against funding for medical research, weapons systems, homeland security, education, and numerous other public programs. The knowledge that resources are limited and trade-offs will be made contributes to a competitive rather than a collaborative environment for facilities investment decision making at all steps in the process. The current federal operating environment may be characterized by guarded communication about facilities investments, adversarial relationships, and gamesmanship.

The Annual Budget Process and Procedures

It is standard practice for private-sector organizations to make decisions about operating and capital expenditures (e.g., facilities) and to budget for them separately; the two are linked through an overall management plan. In the federal government, expenditures for operating and capital expenditures are considered concurrently, and decision making for facilities investments is driven in large part by the annual budget process. The budget scorekeeping rules mandated as part of the Budget Enforcement Act of 1990 (for which there is no private-sector analogue) also influence decisions related to the acquisition or leasing of facilities and the use of alternative financing approaches. The budget process and scorekeeping procedures reinforce a focus on the short term and on the first costs of facilities investments, typically only 5-10 percent of the total life-cycle costs.

An additional complication is that

> [n]early every federal agency oversees some capital spending. . . . As a result, decisions on infrastructure are largely ad hoc in that they are aligned with agencies' programs, which have differing goals. Even within agencies with significant infrastructure budgets like the Department of Transportation, infrastructure investment strategies for different programs like transit and aviation may be developed separately. Because the federal government does not have an overall plan for its capital investments, the challenge of selecting the most important or

cost-effective projects is even more difficult across federal agencies (GAO, 2000b, p. 44).

Procedures

In the federal government, activities typically related to facilities investment and management, such as budgeting, acquisition, and modifying a project's scope or direction to account for changes in requirements, are procedurally encumbered. The Federal Property and Administrative Services Act of 1949 governs the administration of facilities in all federal civilian and military departments and agencies. More recent legislation, including the Government Performance and Results Act of 1993, the Federal Acquisition Streamlining Act of 1995, and the Clinger-Cohen Act of 1996, applies to facilities management but also to a wide range of other federal activities.[1]

The inherent differences between nongovernmental and governmental organizations, then, are significant. Nonetheless these differences do not fundamentally change the need to apply best practice principles and policies to foster successful investment in and management of federal facilities portfolios. They do, however, impact the particular lessons that might be transferred from one domain to the other. The next section focuses on answering the question, How can the principles and policies used by best-practice organizations be applied to the federal operating environment?

ADAPTING BEST-PRACTICE PRINCIPLES AND POLICIES TO THE FEDERAL ENVIRONMENT

Principle/Policy 1. Best-practice organizations establish a framework of procedures, required information, and valuation criteria that aligns the

---

[1] These include the Government Performance and Results Act (GPRA) of 1993 (P.L. 103-62), which requires federal agencies to develop mission statements, long-range strategic goals and objectives, and annual performance plans and to identify and measure the "outcomes" or results of federal programs. Related legislation includes the Chief Financial Officers Act of 1990; the Federal Acquisition Streamlining Act of 1994, Title V; the Government Management Reform Act of 1994; and the Federal Financial Improvement Act of 1996. Executive initiatives and directives that specifically pertain to federal facilities and infrastructure include Executive Order No. 12893, "Principles for Federal Infrastructure Investments" (January 26, 1994); Office of Management and Budget (OMB) Bulletin No. 94–16, Guidance on Executive Order No. 12893, "Principles for Federal Infrastructure Investments"; OMB Circular A–11: Part 3: "Planning, Budgeting, and Acquisition of Capital Assets"; OMB's Capital Programming Guide, a Supplement to Part 3; and Executive Order No. 13327, "Federal Real Property Asset Management," signed February 4, 2004. The President's Management Agenda, issued in the summer of 2001, focuses on improving the management and performance of the federal government.

goals, objectives, and values of their individual decision-making and operating groups to achieve the organization's overall mission; create an effective decision-making environment; and provide a basis for measuring and improving the outcomes of facilities investments. The components of the framework are understood and used by all leadership and management levels.

Discussion 1. The components of this framework include terminology that is agreed upon by the relevant decision-making and operating groups; a business case analysis; evaluation processes that are clearly defined and incorporate multiple decision points; performance measures; continuous feedback processes; methods for establishing accountability; and incentives for groups and individuals.

In the federal government, decisions about federal facilities investments involve multiple stakeholders: Congress and its various committees, the administration, federal departments and agencies that own facilities, operating groups that manage facilities portfolios, the OMB, agencies that use facilities provided by others, special interest constituencies, the GAO, and others. These stakeholder groups have differing terminologies, responsibilities, objectives, and values.

For example, groups that manage facilities portfolios are responsible for ensuring that facilities perform well enough to support their department's or agency's missions and programs without undue disruption. They have limited authority to determine what investments are made within the funding allotted to them. Their objectives and values may be to build the highest-quality facilities within the available budget in order to minimize long-term building operating costs.

Senior-level executives, in contrast, are responsible for the overall performance of the organization in meeting its mission and for using resources effectively and efficiently. They must balance the competing demands of a variety of programs and initiatives: Funding for facilities investments must be weighed against funding for personnel, information technologies, research, other physical assets such as vehicles, ships, planes, and so forth. Their objectives and values may support building a less costly facility of sufficient quality to meet only the immediate need so that investments in other programs can also be made.

Personnel at OMB are responsible for reviewing the budgets submitted by agencies and recommending resource allocations, although they do not make final decisions. Their objectives may include helping to reduce the budget by limiting funding levels for various programs or services. They may not support allocating any funding for building a specific facility.

Decision makers in Congress and the President are asked to balance the competing demands of very different programs across a wide spectrum of agencies and other federal entities: Funding for facilities investments must be weighed against funding for medical research, weapons systems, homeland security, education, or any of a myriad of other public services. At this level, specific facilities investment decisions may be subsumed entirely by policy decisions.

The lack of alignment in goals and objectives among these stakeholders is exacerbated by the federal budget process. The knowledge that resources are limited and trade-offs will be made contributes to a competitive rather than a cooperative decision-making environment. Agencies may overstate a need in their budget requests based on an expectation that the budget will be cut; then, when cuts are made, there may still be enough funding to proceed. Reviewing authorities, in turn, may suspect that budget requests are always inflated and that cuts can safely be made.

A history of such gamesmanship sows elements of doubt and mistrust between the managers providing information and the decision makers using it: Can decision makers believe what is being communicated? Will people do what they promise? Agency managers may believe that decision makers are unable to acknowledge the legitimacy of the needs being set forth in the face of pressure (such as to maintain a certain level of budget request). Or, based on their objectives and values, decision makers may recognize the needs but believe that other investments have a higher priority.

For these reasons and others, the environment for decision making about federal facilities investments can presently be characterized as one of adversarial relationships, gamesmanship, miscommunication, and mistrust.

The committee believes that a framework of procedures, required information, and valuation criteria based on the principles and policies used by best-practice organizations for facilities investment and management should be adopted. The individual missions, goals, cultures, and organizational structures of federal departments and agencies can be expected to result in varying practices within this to-be-adapted government-wide framework of principles and policies.

Because such a framework represents a significant departure from current operating procedures, it might be advisable to establish one or more pilot projects. A small government agency with a diverse portfolio of facilities might provide a good environment in which to test the framework.

RECOMMENDATION 1. The federal government should adopt a framework of procedures, required information, and valuation criteria for federal facilities investment decision making and management that incorporates all of the principles and policies enumerated by this committee.

******

Principle/Policy 2. Best-practice organizations implement a systematic facilities asset management approach that allows for a broad-based understanding of the condition and functionality of their facilities portfolios—as distinct from their individual projects—in relation to their or-

ganizational missions. **Best-practice organizations ensure that their facilities and infrastructure managers possess both the technical expertise and the financial analysis skills to implement a portfolio-based approach.**

Discussion 2(a). Facilities portfolio managers within federal agencies face many challenges, including the following:

1. Finding ways to manage portfolios comprising a few hundred to several hundred thousand individual structures of various types, ages, and conditions without having the authority or budget necessary for proper management. Such portfolios typically are dispersed throughout the United States and sometimes across the world.

2. Coordinating and monitoring several hundred to several thousand ongoing projects for new construction, renovation, repair, and renewal. These projects are in various phases of development and their total costs range from several million to several billion dollars.

3. Adapting 20- to 100-year-old facilities to accommodate new information technologies and new physical security measures.

4. The continued deterioration of facilities as indicated by the growing backlog of maintenance and repair.

5. The acquisition of new facilities without adequate annual resources committed to properly maintain them.

6. Excess and obsolete facilities that consume resources needed for mission-critical facilities or other programs.

In recent years, federal departments and agencies, including but not limited to the Department of Transportation, the Coast Guard, and the GSA, have begun to implement facilities asset management programs that consider both the portfolio and individual investments. Portfolio-based approaches should be implemented in every department and agency with responsibility for facilities management.

RECOMMENDATION 2(a). Each federal department and agency should update its facilities asset management program to enable it to make investment and management decisions about individual projects relative to its entire portfolio of facilities.

Discussion 2(b). A concern in implementing new approaches to facilities asset management is the availability of federal staff with the full range of skills now required. Most federal facilities management organizations currently have facility professionals and staff with expertise in managing contracts, budgets, and schedules related to their specialty. The best of these have also taught themselves communication skills and techniques of financial analysis and information technologies. They have largely done a remarkable job with the resources available to them.

Departments and agencies will need to give their facilities asset·managers training in the business tools and financial theories and concepts required to implement a portfolio-based approach. Mirroring this, departments and agencies, including the OMB and the GAO, should ensure that financial, budget, and program analysts receive some basic training on the physical aspects, not merely the financial aspects, of facilities investment and management.

Training can occur through coursework, seminars in conjunction with the various operating units at the headquarters of an organization, and with operating units in the field. Rotational assignments should be encouraged to provide more in-depth training and understanding. As job vacancies occur in facilities management operating groups, departments and agencies should seek to recruit and hire staff not only with the traditional technical competencies but also with the requisite business-related training.[2]

RECOMMENDATION 2(b). Each federal department and agency should ensure it has the requisite technical and business skills to implement a facilities asset management approach by providing specialized training for its incumbent facilities asset management staff and by recruiting individuals with these skills.

Discussion 2(c). One of the objectives to be met by implementing a facilities asset management approach is to ensure the alignment of an organization's portfolio of facilities with its missions and operating objectives. Continual monitoring is required to identify facilities that are no longer needed due to changing requirements and those that are obsolete technologically or otherwise. Private-sector organizations have a direct incentive to dispose of unneeded facilities as soon as possible because they are a drain on organizational resources and are readily identifiable on their balance sheets. They dispose of excess facilities through sales, nonrenewal or breaking of leases, or demolition to free up resources that can be used for other requirements.

The federal government, in contrast, has continuously acquired facilities over several centuries but placed relatively little emphasis on disposing of facilities that have become obsolete, too costly to maintain, or that do not support current missions and requirements.[3] In some cases, considerable pressure has been placed

---

[2] In Chapter 2 the committee identified numerous institutions that offer the recommended coursework.

[3] There are, of course, federal facilities that are excess but present significant challenges for disposition, such as former nuclear sites and their associated facilities. Clearly such properties must remain under government ownership. Decommissioning such sites will cost billions of dollars; the decommissioning of former uranium enrichment facilities, for example, will cost between $9 billion and $20 billion (NRC, 1996b).

on elected officials by local constituencies to obstruct the closing of local federal facilities, even when it is not economically efficient to continue their use.

Federal policies, practices, and procedures present other obstacles for facilities disposition. Eighty-one separate policies applicable to the disposal of federal facilities have been identified (GSA, 1997). These policies include legislative mandates or directives that are agency-specific as well as government-wide socioeconomic and environmental policies such as the Land and Water Conservation Fund Act of 1965, the Stewart B. McKinney Homeless Assistance Act of 1987, and various historic preservation statutes.

The budget structure also weighs against disposal of unneeded facilities. The budget appropriation line item Operations is used to fund the maintenance, repair, and disposal of facilities for most departments and agencies. Disposal of some excess facilities can occur through transfer of ownership or demolition; in both cases, an up-front investment is required before disposition can occur. Transferring the ownership of a federal facility to a nonfederal entity brings with it the responsibility to meet environmental and other regulations. Depending on the age, materials, and former use of a facility, it may or may not be cost-effective to make the repairs necessary to comply with regulations in order to dispose of it. Similarly, it can be expensive to demolish facilities in the short term even if the long-term benefits may be significant. The military services estimate that demolition costs for facilities range from $8 to $12 per square foot. For the Army alone, demolition of excess facilities could cost more than $1.3 billion (GAO, 1997). Faced with the trade-off of using the available funds to invest in facilities that support current missions or to dispose of excess ones, managers typically choose the first alternative (NRC, 1998).

Finally, there are few incentives for departments and agencies to invest the time and effort to sell excess properties. Proceeds realized through such sales will go to the general treasury, not back to the organization unless it has been given authority under special legislation to retain some portion of them.[4]

RECOMMENDATION 2(c). To facilitate the alignment of each department's and agency's existing facilities portfolios with its missions, Congress and the administration should jointly lead an effort to consolidate and streamline government-wide policies, regulations, and processes related to facilities disposal, which would encourage routine disposal of excess facilities in a timely manner.

Discussion 2(d). Some federal departments and agencies are incurring significant costs for operating and maintaining facilities that they no longer need to support today's missions. The Department of Defense (DoD) estimates it spends

---

[4]Legislation has, in fact, been enacted to allow the U.S. State Department to sell some of its excess properties at fair market value and to retain the proceeds for investment in mission-critical facilities.

$3 billion to $4 billion each year maintaining excess facilities (GAO, 2003f). The Departments of Energy, State, and Veterans Affairs, the GSA, and the U.S. Postal Service own considerable numbers of properties that are no longer needed but continue to require investment of resources (GAO, 2003f). These agencies and possibly others are

> incurring significant costs for staff time spent managing the properties and on maintenance, utilities, security, and other building needs . . . [and] the government is needlessly incurring unknown opportunity costs, because these buildings and land could be put to more cost-beneficial uses, exchanged for other needed property, or sold to generate revenue for the government. . . . [Holding excess properties] presents an image of waste and inefficiency that erodes taxpayers' confidence (GAO, 2003f, p. 11).

The lack of alignment between a department's or agency's missions and its facilities portfolio, coupled with the cost of operating and maintaining excess facilities, can require extraordinary measures to effect some improvement, particularly when the goals, objectives, and values of the President, Congress, departments, and agencies may be so different that a compromise cannot be reached in the traditional operating environment. One such extraordinary measure was the base realignment and closure (BRAC) process used to divest facilities owned by DoD.

As early as 1964, the Secretary of Defense announced the need for a major military base closing. Legislative and executive decision-making groups were unable to reach a compromise on such closures for the next 25 years. Following the end of the Cold War, it became clear again that certain facilities and infrastructure designed to support a specific type of military force were no longer needed or financially supportable. In the 1980s, the report of the Grace Commission concluded that closing unnecessary military bases could produce savings of $2 billion annually. Decision makers in the legislative and executive branches agreed that some infrastructure could be closed down without affecting the capacity of the government to provide for national defense. However, neither branch could close down military bases without the approval of the other. Both branches were reluctant to support the closing of specific bases because of the impacts on local economies, employment, the objections of the local electorate, and the implications for individual members of Congress (Goldfein, 1994).

To resolve this impasse, the Base Realignment and Closure Act was enacted in 1988, establishing a decision-making process outside the traditional operating environment. The Act established an independent commission to make the final recommendations for closures and set ground rules for both executive and legislative agencies in terms of their input and its timing. The Act required elected officials to approve or reject a recommended package of base closings as a whole—elected officials were not allowed to remove individual bases from the list or to otherwise amend it. Time for debate was limited and filibusters in Congress were disallowed. This process was used for three rounds of base closures in the 1980s and 1990s and may be used again in 2005. The government as a whole and DoD in particular have 15 years of experience and lessons learned. Such lessons can be used by DoD and other agencies to make adjustments to the process to improve it and adapt it to other departments and agencies as appropriate.

> **RECOMMENDATION 2(d). For departments and agencies with many more facilities than are needed for their missions—the Departments of Defense, Energy, State, and Veterans Affairs, the General Services Administration, and possibly others—Congress and the administration should jointly consider implementing extraordinary measures like the process used for military base realignment and closure (BRAC), modified as required to reflect actual experience with BRAC.**

******

Principle/Policy 3. Best-practice organizations integrate facilities investment decisions into their organizational strategic planning processes. Best-practice organizations evaluate facilities investment proposals as mission enablers rather than solely as costs.

Discussion 3. Federal departments and agencies typically are established to serve specifically defined missions and objectives and to execute programs to achieve them. Throughout their histories, departments and agencies have conducted strategic planning processes aimed at identifying and achieving short-, intermediate-, and long-term objectives. Strategic planning processes have been formalized and their reporting requirements expanded through the Government Performance and Results Act of 1993.

In regard to facilities investments, most federal departments and agencies have not yet linked their strategic planners and finance directors with their facilities management operating groups, nor have they demonstrably integrated facilities investment decision making into their organizational strategic planning processes. Instead, decision making for facilities investments is typically a reactive planning process that has been described as follows:

> It begins with the lowest or low-level units of an organization identifying the deficiencies and threats they face. Then they attempt to return to a preferred earlier state by designing projects intended to reveal the causes of these deficiencies and threats and to remove or suppress them. Next, using cost-benefit analysis, priorities are assigned to projects. Finally, using an estimate of the amount of resources that will be available for work on projects, a set of them is selected starting at the top of the priority list, working down until all the expected resources have been allocated. The set of projects thus selected constitutes the unit's plan.
>
> Unit plans are passed up to the next higher-level unit, where they are edited and coordinated and integrated with a plan similarly prepared at that unit. This pro-

cess is continued until the accumulated plans reach the top of the organization, where again they are edited, coordinated, and integrated with projects designed at that level. (Ackoff, 1999, p. 104)

Most initiatives or activities contemplated in any department's or agency's organizational strategic planning entail a facilities requirement: Space is required to house people and equipment and to ensure that operations are ongoing and efficient. The location of that space can help or hinder operations. When departments and agencies plan for their facilities investments using a reactive rather than a more integrated management approach, they fail to account for a potentially substantial portion of the total cost of a program or initiative.

Integrating facilities considerations into organizational strategic planning processes up front will provide decision makers with better information about the total long-term costs, considerations, and consequences of a particular course of action. One method for doing so is to have the senior facilities program manager participate in the organization's strategic planning sessions and processes. His or her role is to translate between the organization's mission and programs and its physical assets and to clearly communicate the potential support that enabling real estate and facilities can give to the organization's mission.

Some departments and agencies have already instituted more integrated management approaches to planning. In the State Department, for example, the director/chief operating officer of the Bureau of Overseas Buildings Operations participates in strategic planning sessions with the other senior-level department executives (undersecretaries and assistant secretaries for the various operating units). His role is to link investments in embassies, consulates, and other facilities and the abandonment of still others to the conduct of foreign policy and to help identify the impacts, costs, potential consequences, and opportunities of such investments.

In response to criticism from the OMB and others about its facilities planning and management processes, the VA has developed a planning process that considers trade-offs among all types of physical assets, including infrastructure projects, nonmedical equipment, leases (new and existing of more than 300,000 square feet), medical equipment, information technology, and enhanced-use leases (public-private partnerships) (VA, 2003). The process requires that capital investment proposals be clearly tied to the department's goals and objectives before they are considered for funding. A strategic review is conducted by a capital investment board (CIB) made up of senior executives from the major operating units. The CIB is responsible for evaluating, prioritizing, and measuring proposals against the VA's strategic plan and OMB's requirements. The committee believes that all federal departments and agencies should integrate facilities investment decisions into their organizational strategic planning processes.

RECOMMENDATION 3. Each federal department and agency should use its organizational mission as guidance for facilities investment decisions and should then integrate facilities investments into its organizational strategic planning processes. Facilities investments should be evaluated as mission enablers, not solely as costs.

******

Principle/Policy 4. Best-practice organizations use business case analyses to rigorously evaluate major facilities investment proposals and to make transparent a proposal's underlying assumptions; the alternatives considered; a full range of costs and benefits; and the potential consequences for their organizations.

Discussion 4(a). A business case analysis, as used in the private sector, is not a budget or accounting document, nor is it a static, one-time-only analysis that looks solely at physical assets. It is, instead, a planning and decision-support tool that is constantly revised to reflect changing requirements and to incorporate better or updated information. It accounts for the life-cycle costs of all of the resources inherent in an investment decision—that is, facilities, staff, equipment, technologies, and financial resources.

Federal efforts to provide more complete analyses of facilities investment alternatives have been initiated. The OMB has issued the *Capital Programming Guide,* which incorporates policies and procedures for developing and evaluating alternatives to be used by all executive branch agencies when preparing budget requests. It is intended to provide guidance for a disciplined capital programming process to ensure that capital assets contribute to the achievement of agency strategic goals and objectives (OMB, 1997).[5]

Federal departments and agencies have also issued internal guidance for developing their own business case analyses for facilities investments. As one example only, the VA developed the *Capital Investment Methodology Guide* as a basic reference to help standardize the methods for gathering, analyzing, and presenting data to decision makers.[6] The guide incorporates tools to analyze a proposal's cost-effectiveness, alternatives, risk, and earned value.

The continual updating of a business case analysis is an important consideration for federal departments and agencies, where facilities investment proposals may take years to move through the budgeting process. Private-sector organizations invest minimal resources at the earliest stages of proposal evaluation and

---

[5]Because OMB defines capital assets as "land, structures, equipment and intellectual property (including software) that have an estimated useful life of two years or more," the guide applies to capital assets that are substantially different in character, purpose, and longevity.

[6]Available at www.va.gov/budget/capital.

analysis, focusing primarily on the financial aspects. If the expected return is not sufficient to justify additional study, the proposal is terminated. If additional study is justified, it is undertaken in an iterative manner such that significant resources are only expended once a proposal becomes a project.

In contrast, federal departments and agencies may invest significant resources in conducting an alternatives analysis and conceptual planning, in some cases taking a project to a 35 per cent design phase before a proposal is presented to OMB or Congress. At this point, several hundred thousand dollars or more, and many hours of staff time, will have been expended. As a project receives conditional approvals within the agency itself, from OMB, and from Congress, which may take years, more people and operating groups become vested in the proposal. Most department or agency managers are reluctant to reevaluate the need for a specific project even if it is clear that requirements have changed, because of the buy-in by other groups. Federal managers also take political and financial risks if they request that Congress reallocate an appropriation to another use. Once a project is approved, it is not usually put on a fast track. More commonly, several years will pass before it is actually constructed and occupied.

Continual updating of information may help to preclude the building of facilities designed for a requirement that has been overtaken by events. For example, to mitigate problems caused by an extended time frame, the VA has instituted a policy that after proposals have been approved and funded but before initiation, proposal teams must submit progress reports to determine if schedules and costs would still be on target and must take corrective actions as appropriate. Once funding is secured, planning assumptions approved 18 to 24 months earlier must be reviewed and validated before the obligation of funds.

Because a business case analysis is tailored to the vocabulary, culture, resources, and mission of an organization, it is developed and revised over time and through repeated use by all of the decision-making and operating groups. Thus, there is no standard format for a business case analysis that can be readily adapted to all federal departments and agencies. However, the committee believes that such an analysis can and should be developed by each federal department and agency and refined over time through repeated, consistent use by all of their decision-making and operating groups.

Whatever its format, a federally adapted business case analysis should explicitly include and clearly state the following: (1) the organization's mission, (2) the basis for the requirement for the facility investment, (3) the objectives to be met by the facility investment and its potential effect on the entire facilities portfolio; (4) performance measures for each objective to indicate how well objectives will have been met, (5) identification and analysis of a full range of facilities investment and other alternatives to meet the objectives, including the alternative of no action, (6) descriptions of the data, information, and judgments necessary to describe anticipated performance of the alternatives in terms of performance measures, (7) a list of the value judgments (i.e., value trade-offs) made to balance achievement on competing objectives, (8) a logic for the overall evaluation of the alternatives, (9) strategies for exiting the investment, and (10) the names of the individuals and operating groups responsible for the analysis and accountable for subsequent performance. The form of the business case analysis for each department and agency should be agreed to by the appropriate oversight authorities.

RECOMMENDATION 4(a). Each federal department and agency should develop and use a business case analysis for all significant facilities investment proposals to make clear the underlying assumptions, the alternatives considered, the full range of costs and benefits, and the potential consequences for the organization and its missions.

Discussion 4(b). One element of the recommended framework is common terminology to promote effective communication among the various stakeholders when discussing the business case analysis or other facilities-related issues. Engineers, lawyers, accountants, economists, technologists, military personnel, and elected officials lack a common vocabulary or style of interaction. Nor do facility planners, facility operators, agency heads, facility users, legislative personnel, budget analysts, and elected officials necessarily share a common set of interests or time frames they consider important. Common terminology promotes improved communications among stakeholders with widely differing educational and technical backgrounds, values, and objectives.

For effective communication to occur, facilities asset management staff should have the capacity and skills to understand the relationship of facilities to the big picture—that is, the organizational mission—and to communicate that understanding. They should also be able to solve problems by considering all sides of issues and negotiate a solution that will best meet the organizational requirement. Similarly, the staff of reviewing authorities should have the capacity and skills to understand the physical aspects of facilities management as practiced in the field. Training, rotational assignments, and cultivating a wide variety of contacts and relationships through networking are effective methods for instilling such skills.

RECOMMENDATION 4(b). To promote more effective communication and understanding, each federal department and agency should develop a common terminology agreed upon with its oversight constituencies for use in facilities investment deliberations. In addition, each should train its asset management staff to effectively communicate with groups such as congressional committees having widely different sets of objectives and values. Mirroring this, oversight constituencies should have the capacity and skills to understand the physical aspects of facilities management as practiced in the field.

******

Principle/Policy 5. Best practice organizations analyze the life-cycle costs of the proposed facilities, the life-cycle costs of staffing and equipment inherent to the proposal, and the life-cycle costs of the required funding.

Discussion 5(a). In the federal government, policies and directives for using life-cycle costing of facilities investments have been issued. However, because capital and operating expenditures are considered concurrently, the annual budget process does not encourage a total life-cycle perspective at the highest levels of decision making.

Under the current budget structure, only the projected design and construction (first costs)—which account for only 5 to 10 percent of the total costs of facilities—are easily identifiable and open to scrutiny by the OMB, Congress, and others. Funding requests for design and construction are considered on a case-by-case basis under separate line items. In contrast, funding requests for the operation, maintenance, and disposal of facilities are lumped together in a different line item and may be considered in different budget years. Thus, the budget process is so structured that up to 95 percent of the total life-cycle costs of operating and maintaining facilities are not routinely considered.

For some high-dollar project proposals, federal departments and agencies conduct life-cycle analyses internally to understand the total facilities costs and benefits over the long term and to prioritize their requests for funding. However, once their budget requests are submitted, the requests are disaggregated into funding for design, construction, operations, and maintenance of the facility to conform to the budget structure; the full costs of staffing, equipment, and technologies for the particular facility are not included. In its research and interviews, the committee was not made aware of any instance in which a department or agency conducted a life-cycle analysis for a facility investment proposal *and* a life-cycle analysis of its attendant staffing, equipment and technologies *and* a life-cycle analysis of the cost of funding. If agencies were to adopt more integrated management and planning approaches, such analyses would probably become more commonplace.

RECOMMENDATION 5(a). Each federal department and agency should use life-cycle costing for all significant facilities investment decisions to better inform decision makers about the full costs of a proposed investment. A life-cycle cost analysis should be completed for (1) a full range of facilities investment alternatives, (2) the staff, equipment, and technologies inherent to the alternatives, and (3) the costs of the required funding.

Discussion 5(b). The focus on the first costs of facilities investments is reinforced by the budget scorekeeping rules mandated as part of the Budget Enforcement Act of 1990, discussed in Chapter 5. Revising the budget scorekeeping rules such that they meet congressional oversight objectives for transparency and at the same time facilitate decision making that takes into account the long-term interests of departments and agencies as well as the public will not be an easy task. Amending the scorekeeping rules specifically to account only for life-cycle costs would probably create an even greater disincentive for facilities investments. The committee believes that a collaborative effort that encompasses a wide range of objectives, goals, and values is required. Some possible revisions to the rules could be tested in pilot projects.

RECOMMENDATION 5(b). Congress and the administration should jointly lead an effort to revise the budget scorekeeping rules to support facilities investments that are cost-effective in the long term and recognize a full range of costs and benefits, both quantitative and qualitative.

******

Principle/Policy 6. Best-practice organizations evaluate ways to disengage from, or exit, facilities investments as part of the business case analysis and include disposal costs in the facilities life-cycle cost to help select the best solution to meet the requirement.

Discussion 6. When planning for new facilities or major renovations, federal departments and agencies typically do not develop exit strategies.[7] When considering the acquisition of new facilities, it is not yet commonplace to analyze the entire portfolio of facilities to determine whether other existing facilities will become obsolete to the mission or to analyze the resulting cost implications.

The development of exit strategies as part of a business case analysis will help federal decision makers to better understand the potential consequences of the alternative approaches. An evaluation of exit strategies can, for example, provide a basis for determining whether it is best to own or lease the required space in a particular situation, or whether specialized or more generic flexible space is the best solution. For those investment proposals in which the only exit strategy is demolition and cleanup, evaluating the costs of disposal may lead to better decisions about the design of the facility and the choice of materials, thereby reducing life-cycle costs.

RECOMMENDATION 6. Every major facility proposal should include the strategy and costs for exiting the investment as part of its business case analysis. The development and evaluation of exit strategies during

---

[7]An example of an exit strategy is housing vouchers that allow enlisted men and women to seek housing in the private market. Using vouchers provides the DoD and its military services with more flexibility to adjust to fluctuations in staff needs and allows them to avoid the long-term costs and commitments of operating military housing.

the programming process will provide insight into the potential long-term consequences for the organization, help to identify ways to mitigate the consequences, and help to reduce life-cycle costs.

******

Principle/Policy 7. Best-practice organizations base decisions to own or lease facilities on the level of control required and on the planning horizon for the function, which may or may not be the same as the life of the facility.

Discussion 7. As do private-sector organizations, federal departments and agencies acquire the use of space and equipment through ownership and through operating and capital leases. Based on the committee's interviews and research activities, the criteria departments and agencies use to determine if it is more cost-effective to own or lease facilities to support a given function are not clear or uniform. What *is* clear is that the decision is complicated by the budget scorekeeping rules.

As with budget requests to design and construct facilities, requests to fund operating and capital leases are scored up front.[8] Leases typically will have lower costs over the given lease period than the design and construction of a facility, even if the long-term costs might be higher. Consequently, leasing the required space may appear to have less impact on an organization's overall budget. In this case, the scorekeeping rules may provide an incentive for a department or agency to game the system and request approval to lease space even if it is not cost-effective in the long term.

The committee believes that a more effective approach for deciding whether it is best to own or lease the required space is to base the decision on the level of control desired and the planning horizon for the function. Departments and agencies should determine whether the required space will support functions that are critical to the organizational mission (core functions), functions that support the mission, or functions that are mission neutral (noncore) and then determine the level of control desired. An additional decision factor should be the length of time the function must be supported. For long-term, mission-critical functions, a department or agency may wish to exert maximum control through ownership. For short-term, noncore functions, leasing may be the most cost-effective option. Whatever the decision, the committee believes it should be based on a clearly stated rationale linked to support of the organizational mission and the life of the function.

---

[8]The criteria used to distinguish among the different categories of leases are to some extent arbitrary (CBO, 2003), leading to some variation in the ways leases are scored by the OMB and the CBO.

RECOMMENDATION 7. Each federal department and agency should base its decisions to own or lease facilities on the level of control desired and on the planning horizon for the function, which may not be the same as the life of the facility.

******

Principle/Policy 8. Best-practice organizations use performance measures in conjunction with both periodic and continuous long-term feedback to evaluate the results of facilities investments and to improve the decision-making process itself.

Discussion 8. The Government Performance and Results Act of 1993 requires all federal departments and agencies to develop performance measures in order to evaluate the effectiveness of their programs and investments in providing goods and services to the American public and to report the results annually. Some federal organizations have used the Balanced Scorecard concept to develop measures for determining how well strategic objectives are being met. However, because the results of many federal programs or services are qualitative in nature and occur over long periods of time—for example, the regulation of air quality—measuring them can be challenging.

Federal organizations are faced with several issues when they develop performance measures to capture the outcomes of facilities investments and management as they apply to portfolios of facilities. One is the lack of adequate baseline data about facilities portfolios: their condition, value, functionality, and operating costs. When Congress appropriates the annual operations and maintenance budget back to the agencies, the agencies themselves then allocate this funding to investments in facilities maintenance, repair, alteration, and renewal. (Planning, design, and construction of projects are typically funded through separate line items.) Departments and agencies do not systematically separate and track actual expenditures for maintenance, repair, and operations of buildings, making it difficult to develop accurate baseline data. A second issue is the structure of current accounting systems, which are driven by the federal budget process and do not typically disaggregate facilities operations and maintenance costs. An example of one further complicating factor is that many federal buildings are not metered, making it difficult to track utility costs or usage.

Despite these and other difficulties, efforts are under way to develop measures that apply to facilities portfolios. Many agencies use a facilities condition index (FCI) to monitor the overall condition of their facilities inventories. The Navy has developed the Mission Dependency Index (MDI), which uses operational risk management techniques of probability and severity and applies them to facilities in terms of interruptability, relocatability, and replaceability. The MDI also takes mission intradependencies (those that reside within a command) and mission interdependencies (those that reside between commands) into account through a structured interview process with command representatives of individual units that cover a finite geographical area. The DoD has developed a facilities sustainment model and recapitalization metric to determine the rate of restoration and modernization relative to the average expected service life of the inventory. The Coast Guard is developing (1) a space utilization index (SUI) to measure compliance with organizational space standards to ensure equitable distribution of space and funding across the organizations and (2) a systems criticality index based on functional importance, health and safety, repair cost factors, interdependencies with other systems, and other factors (Dempsey et al., 2003). NASA has developed a parametric model for tracking deferred maintenance across its inventory. Some or all of these indices could be adapted for use by other federal departments and agencies and used in combination with other metrics to measure the performance of federal facilities portfolios. An approach like that of the Balanced Scorecard could be applied hierarchically, with successively more detailed objectives and metrics at lower levels. The department or agency level would be the starting point since it is the focus of resource allocation and establishment of management objectives. Department objectives would flow down to agencies and thence to regions and facilities.

The effective use of performance measures for facilities investment and management requires the continuous monitoring of projects, processes, and facilities portfolios through short- and long-term feedback. Monitoring the progress of facility projects to measure whether they are on time and within budget is a common practice in federal organizations. The GSA, the U.S. Postal Service, the State Department, and other agencies receive feedback on customer satisfaction with newly occupied buildings through postoccupancy evaluations. Such evaluations provide a basis for lessons-learned programs, which in turn are used to improve processes and design standards. However, most of the feedback procedures are short term. To the committee's knowledge, no federal department or agency gathers consistent, organized, long-term feedback to determine if facilities investments met organizational objectives, solved operational problems, or reduced long-term operating costs. This type of feedback is essential if the outcomes of facilities investments and management processes are to be measured and improved.

RECOMMENDATION 8. Each federal department and agency should use performance measures in conjunction with both periodic and continuous long-term feedback and evaluation of investment decisions to monitor and control investments, measure the outcomes of facilities investment decisions, improve decision-making processes, and enhance organizational accountability.

******

Principle/Policy 9. Best-practice organizations link accountability, responsibility, and authority when making and implementing facilities investment decisions.

Discussion 9. As noted in *Stewardship of Federal Facilities: A Proactive Strategy for Managing the Nation's Public Assets*, the responsible ownership of facilities by the federal government is

> an obligation that requires not only money, but also the vision, resolve, experience, and expertise to ensure that resources are allocated effectively to sustain the public's investment. The recognition and acceptance of this obligation is the essence of stewardship. Public officials and employees at all levels are responsible for decisions that affect the stewardship of facilities. (NRC, 1998, p. 62)

In the federal government, responsibility and authority for making decisions and executing programs often are not directly linked. Instead, decision-making authority and decision-making responsibility are spread throughout the executive and legislative branches, leading to a lack of clear-cut accountability for facilities investment outcomes.

In the instance of facilities management and maintenance, the linkages between responsibility, authority, and accountability are lacking at several levels. First, for most facilities projects, one operating unit may be responsible for the planning of a facility, another for designing and constructing it, and a third for maintaining, preserving, and operating it. When these functions are separate, there is no strong incentive for those designing a facility to consider its life-cycle costs or to evaluate alternative materials, systems, or other components in terms of their impact on long-term operations and management, repair, and disposal costs. The groups responsible for design are rarely held accountable for the subsequent total operating costs of the facility. The group overseeing construction is responsible for and held accountable for completing the facility on time and on budget but not necessarily for ensuring that the facility will operate economically and satisfy user requirements.

Second, those who use a facility often are not responsible for its maintenance and care. Their budget allocation is usually separate from that for facilities maintenance, so to them, the facility does not have a direct cost. They operate within the facility but are not accountable for how their operations affect the facility or the cost of maintaining it.

Third, those responsible for managing facilities portfolios may be held accountable for the quantity and quality of services being provided to the organization. However, they are not always given the resources and authority necessary to maintain the facility's functionality or condition at the level needed to effectively support the required services.

Fourth, to help balance the budget, budget and program analysts may be responsible for limiting the resources to be invested. However, they are not held accountable for the consequences of their recommendations, which may include the worsening condition of facilities over time through lack of investment.

Fifth, senior managers and elected officials are responsible for broad levels of services and for balancing needs with available resources. At this level, trade-offs are made among a wide range of programs and services, and decisions are made by consensus. Faced with these trade-offs, senior managers and public officials may decide that there will be no serious consequences if facility maintenance and repair is deferred another year in favor of more urgent operations or programs with greater visibility. Only if there is a catastrophic failure, such as a roof falling in or a bridge collapsing, are senior managers likely to be held accountable for the condition of facilities in any given year (NRC, 1998).

Within the federal government, private-sector methods for linking responsibility, authority, and accountability for facilities investment-related activities are most easily applied at the project level. Given adequate resources and the authority to allocate those resources, a facility project manager can be held accountable for delivering a project on time and within budget. As one moves up within a department or agency to the facility program level, accountability for the outcomes of investments is more difficult to establish owing to the typical separation of planning, design, construction, and operations functions and, more importantly, to an inability to control adequate resources to manage existing facilities portfolios over the long term.

Several elements of the framework of principles and policies recommended in this report will enhance accountability for the outcomes of federal facilities investments. Implementation of facilities asset management approaches, coupled with adequate resources and authority for allocating those resources, will enhance accountability for outcomes within facilities management organizations. A facilities asset management approach allows linking the performance of the facilities portfolio to the organization's mission and measuring how well operational and strategic objectives are being met over both the short and long terms.

The development and consistent use of a business case analysis that documents decisions, value trade-offs, the quality and depth of the alternatives analyzed, and those responsible for the analysis will enhance accountability for investment proposals and their outcomes. More integrated approaches to the design, construction, and operation of individual buildings could result in lower life-cycle costs and could also serve to make planners, designers, constructors, and operators of facilities more accountable for the performance and functionality of the facility. Some design-build-operate-maintain project delivery strategies have been developed on these premises. The Departments of Defense and State are now conducting pilot studies to determine if this type of project delivery strategy could be widely used to achieve better facilities and greater accountability.

As a first step toward making the decision-making process itself more transparent, and to enhance accountability at all levels, each federal department and agency should develop a decision tree or diagram that illustrates the many interfaces among the decision-making and operating groups involved in the process, identifies the points at which decisions are made, and identifies the groups making the decisions at each point.

RECOMMENDATION 9. To increase the transparency of its decision-making process and to enhance accountability, each federal department and agency should develop a decision process diagram that illustrates the many interfaces and points at which decisions about facilities investments are made and the parties responsible for those decisions. Implementation of facilities asset management approaches and consistent use of business case analyses will further enhance organizational accountability.

******

Principle/Policy 10. Best-practice organizations motivate employees as individuals and as groups to meet or exceed accepted levels of performance by establishing incentives that encourage effective decision making and reward extraordinary performance.

Discussion 10. The federal government, unlike private-sector organizations, does not operate on a risk-reward basis, nor does it seek to make a profit. Using public dollars to create financial incentives to motivate individuals to meet organizational objectives sometimes raises concerns; however, such incentives are already used on a limited basis by federal departments and agencies. Incentives come in many forms. Identifying and implementing incentives to support good decision making on the part of individuals and operating groups is as important for federal organizations as it is for the private sector.

In the federal system, the multiple-objective nature of laws and policies and the sheer volume of procedures sometimes result in unintended consequences, including disincentives for good decision making and cost-effective behavior. The budget scorekeeping rules are one example; they are intended to provide transparency in the budget and to help control spending, but they also engender gamesmanship that discourages long-term, cost-effective behavior in favor of behavior that satisfies short-term needs. The separation of planning, design, construction, and operations functions within departments and agencies creates disincentives for life-cycle costing in favor of driving down the first costs of facilities. The federal budget process creates additional disincentives for cost-effective actions. For example, in most circumstances, the carryover of unobligated funds from one fiscal year to the next is not allowed even if a facilities program manager can demonstrate that carryover of funding for a capital investment is the most cost-effective approach. Funds that are not expended in the current fiscal year are routinely taken back from departments and agencies, and the next fiscal year's funding may be reduced on the premise that money not spent is money not needed. Thus, admitting to savings is not in a federal manager's interest or that of his organization (NRC, 1998).

Examples of incentives that would support more cost-effective decision making and management by facilities asset management groups include these: (1) allow savings from one area of operations to be applied to needs in another area if the savings are carefully documented; (2) allow the carryover of unobligated funds from one fiscal year to the next for capital improvements, if doing so can be shown to be cost-effective; or (3) establish awards for operating units with high levels of performance. A major issue in the implementation of such programs is to find ways to militate against the common practice of reducing a department's or agency's budget in future fiscal years if the agency appears to have funds available at the end of the current year.

GSA's Public Buildings Service (PBS) has instituted a program for linking budget to performance that provides one example of how financial incentives can be applied in the federal government to motivate operating groups to better meet organizational goals on a national and regional level. In 1998, PBS began using a limited number of performance metrics and targets, coupled with funds from its annual budget, to precipitate changes in employee performance. The funds for the program are set aside at the beginning of each fiscal year. For each of the nine performance measures, which are organized around business performance and customer satisfaction, PBS leadership sets a national goal. It then negotiates targets for each of its 11 regions, taking into account the characteristics of the real estate markets in each region. At the end of the fiscal year, the funds are distributed to those regions that meet or exceed the national goal; the regions that do not meet goals do not share in the bonus pool. Regions have discretion in how the money is used. Of $75 million distributed as of 2001, approximately two-thirds was used for repairs and alterations of PBS space to improve long-term performance of regional facilities inventories and about one-third for salary, training, workspace, and team awards (Dunham and Beard, 2001). The PBS also reports improved collaboration among the regions and significant improvements in their performance as additional outcomes of the program.

RECOMMENDATION 10. Congress and the administration and federal departments and agencies should institute appropriate incentives to reward operating units and individuals who develop and use innovative and cost-effective strategies, procedures, or programs for facilities asset management.

AN OVERALL STRATEGY FOR IMPLEMENTATION

Transforming decision-making processes, outcomes, and the decision-making environment for federal facilities investments will require sponsorship, leadership, and a commitment of time and resources from many people at all levels of government and from some people outside the government. Implementation of some of the committee's recommendations can begin immediately within federal departments and agencies that invest in and manage significant portfolios of facilities. However, implementing an overall framework of principles and policies will require collaborative, continuing, and concerted efforts among the various legislative and executive branch decision makers and operating groups. These include the President and Congress, senior departmental and agency executives, facilities program managers, operations staff, and budget and management analysts within departments and agencies and from the CBO, the OMB, and the GAO.

Having noted this, the committee is well aware that similar recommendations made by other learned panels advocating long-term, life-cycle stewardship of facilities and infrastructure have achieved only limited success (see, for example, NCPWI, 1988; NRC, 1990, 1991, 1998) and have failed to motivate those outside the professional facilities community to action. The committee believes that a new dynamic can and must be instituted.

An illustrative model of sociotechnical systems (Figure 6.1) is useful for visualizing the interactions that occur during a complex decision-making process (Linstone, 1984). If the committee's recommendations for improved decision making for federal facilities investments are to be implemented successfully, these interactions must be understood and enabled by all the participants in federal facilities investments and management. Facility managers will not be successful if they limit themselves to narrow technical analyses or if interactions with senior

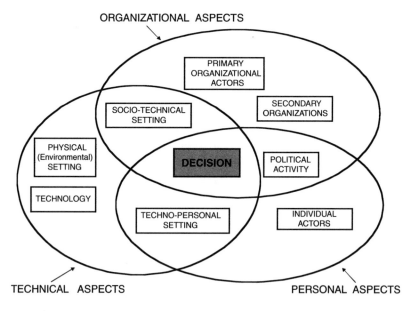

FIGURE 6.1 A sociotechnical system view for decision making. SOURCE: Linstone, 1984

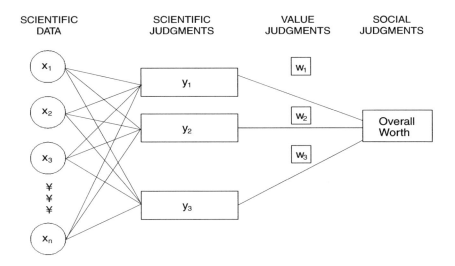

FIGURE 6.2 A model for integrating scientific and social values in decision making. SOURCE: Hammond, 1996.

agency management, program and financial staff, and OMB occur just once a year as part of the budget cycle. Building a case for proactive facility investments requires that dialogue be initiated and sustained between and among the various stakeholders using terms of reference that all can relate to and act upon.

Kenneth Hammond has researched the issue of integrating scientific and social values into the decision-making process and applied the results in practical ways (Hammond, 1996). He found that in public decision-making settings, an impasse may occur despite numerous meetings and discussions between government officials and community leaders. Often, the root cause of the impasse is the fact some stakeholders are concentrating solely on technical factors, while community leaders are primarily concerned with the potential effects on the citizenry.

Figure 6.2 is a diagram Dr. Hammond developed to demonstrate that the various stakeholders often address related but distinct problems: the technical requirements involved in solving a specific issue and the social values of the community. Once such differences are recognized, discussions can be shaped to address a full range of issues and to develop trust and understanding among the stakeholders.

The committee believes that all too often, the facilities management community, whether in the public or private sector, presents an analysis designed to convince those who already believe in good facility practices. Factors from the left side of Figure 6.2 are developed and honed to a keen edge. However, this information often fails to sway the decision makers, who count facilities as only one area among many competing for resources and attention. Demonstrating that proactive facility investment supports the broader values of the organization or government entity will allow for integrated decision making that is more compelling to all stakeholders.

Implementing a framework of expectations, processes, information, and criteria based on the principles and policies identified by the committee will require broad sponsorship, focused leadership, and deep commitment on the part of all stakeholders.

To this end, the committee recommends that legislation be enacted and executive orders be issued that would do two things:

(1) Establish an executive-level commission with representatives from the private sector, academia, and the ranks of the federal government to determine how the identified principles and policies can be applied in the federal government to improve the outcomes of decision-making and management processes for federal facilities investments within a time certain. The executive-level commission should include representatives from nonfederal organizations acknowledged as leaders in managing large organizations, finance, engineering, facilities asset management, and other appropriate areas. The commission should also include representatives of Congress, federal agencies with large portfolios of facilities, oversight agencies, and others as appropriate. It should be tasked to gather relevant information from inside and outside the federal government; hold public hearings; submit a report to the President and Congress outlining its recommendations for change; an implementation plan; and a feedback process for measuring, monitoring, and reporting on the results—*all within a time certain.*

(2) Concurrently establish department and agency working groups to provide recommendations to the executive-level commission for use in its deliberations. The working groups within each department and agency should work collaboratively with the executive-level commission. Staff in the departments and agencies are in the best position to communicate their organizational culture and identify practices for implementing the principles and policies that will work for their organization. In addition, they can provide the commission with information on the characteristics of their facilities portfolios; issues related to aligning their portfolios with their missions; facilities investment trends; good or best practices for facilities investment and management; performance measures for monitoring and measuring the results of investments; and other relevant information.

The committee believes that such sponsorship, leadership, and commitment to this effort will result in

- Improved alignment between federal facilities portfolios and missions to better support our nation's goals,
- Responsible stewardship of federal facilities and federal funds,
- Substantial savings in facilities investments and life-cycle costs,
- Better use of available resources—people, facilities, and funding,
- Creation of a collaborative environment for federal facilities investment decision making.

Bibliography

Ackoff, R.I. 1999. Ackoff's Best. New York: John Wiley & Sons.

Al-Tamimi, Hussein, A. Hassan, and David VanderLinden. 2001. Capital budgeting practices of manufacturing firms in the UAE. Journal of Emerging Markets 6(1): 47-58.

Amekudzi, Adjo, Pannapa Herabat, Kristen Sanford Benhardt, and Sue McNeil. 2001. Educating students to manage civil infrastructure systems. Proceedings of the 2000 Annual Conference of the American Society for Engineering Education (ASEE). Washington, D.C.: ASEE.

American Public Works Association (APWA), Task Force on Asset Management. 1998. Asset Management for the Public Works Manager: Challenges and Strategies. Available at http://www.apwa.net/documents/resourcecenter/ampaperv2.doc.

Association of Higher Education Facilities Officers (APPA). 2000. Strategic Assessment Model, 2nd ed. Alexandria, Va.: APPA.

Block, Stanley. 2000. Integrating traditional capital budgeting concepts into an international decision-making environment. The Engineering Economist 45(4): 309-325.

Blunt, Ray. 2001. Organizations Growing Leaders: Best Practices and Principles in the Public Service. The PricewaterhouseCoopers Endowment for the Business of Government. Washington, D.C.: Council for Excellence in Government.

Borins, Sandford. 2001. The Challenge of Innovating in Government. The PricewaterhouseCoopers Endowment for the Business of Government. Arlington, Va.: PricewaterhouseCoopers Endowment.

Brandt, R.M. 1994. A Guide to Strategic Facilities Planning. Holland, Mich.: Haworth, Inc.

Brill, Michael, S. Margulis, and E. Konar. 1985. Using Office Design to Increase Productivity. Edmond, Okla.: Environmental Design Research Association.

Bureau of Medicine and Surgery (BUMED). 1999. Draft business case analysis. Available at http://nhso.med.navy.mil/BCA.

Campbell, J. 2000. BYU's CNA Center expands the life of its campus. Facility Management Journal (May/June): 52-56.

Catbas, F. Necati, and A. Emin Aktan. 2002. Condition and damage assessment: Issues and some promising indices. Journal of Structural Engineering 128(8): 1026-1036.

CCAF~FCVI. 2002. Reporting Principles: Taking Public Performance Reporting to a New Level. Ottawa, Ontario: CCAF~FCVI Inc.

Center for Construction Industry Studies (CCIS). 1999. Owner/Contractor Organizational Changes, Phase II Report. Austin: University of Texas Press.

Collins, Jim C. 2001. Good to Great: Why Some Companies Make the Leap . . . And Others Don't. New York, N.Y.: Harper Business.

Congressional Budget Office (CBO). 2003. The Budgetary Treatment of Leases and Public/Private Ventures. Washington, D.C.: Congressional Budget Office.

Cotts, D., and E.P. Rondeau. 2004. The Facility Manager's Guide to Finance and Budgeting. New York: AMACOM.

Dempsey, J.J., J. Watson, R. Skulte, W.J. Facsenmeier, and G. Robertson. 2003. SFCAM Metrics—Part 5: A redefined facility condition index. Pp. 8-12 in U.S. Coast Guard Shore Facility Capital Asset Management: Concepts, Strategies, and Metrics. White Paper. Available at http://web.lmi.org/uscg.

Doss, C.B. 1987. The use of capital budgeting procedures in U.S. cities. Public Budgeting and Finance 7(3): 57-69.

Drury, C., and M. Tayles. 1997. The misapplication of capital investment appraisal techniques. Management Decision 35(2): 86-93.

Duckworth, Steven L. 1993. Realizing the strategic dimension of corporate real property through improved planning and control systems. Journal of Real Estate Research 8(4): 495-509.

Dunham, E., and C. Beard. 2001. Presentation/discussion to Balanced Scorecard Interest Group on October 25, 2001. Available at unpan1.un.org/intradoc/groups/public/documents/aspa/unpan001989.pdf.

Eljelly, Abuzar M.A., and Abubakr M. Abuidris. 2001. A survey of capital budgeting techniques in the public and private sectors of a less developed country (LDC): The case of Sudan. Journal of African Business 2(1): 75-93.

Erdener, E. 2003. Linking programming and design with facilities management. Journal of Performance of Constructed Facilities 17(1): 4-8.

Federal Energy Management Program (FEMP). 2000. Annual Report to Congress on Federal Government Energy Management and Conservation Programs. Fiscal Year 1998. Washington, D.C.: FEMP.

FEMP. 2001. Annual Report to Congress on Federal Government Energy Management and Conservation Programs. Fiscal Year 1999. Washington, D.C.: FEMP.

FEMP. 2003a. Annual Report to Congress on Federal Government Energy Management and Conservation Programs. Fiscal Year 2001.Washington, D.C.: FEMP.

FEMP. 2003b. The Business Case for Sustainable Design in Federal Facilities. Washington, D.C.: U.S. Department of Energy.

Federal Facilities Council (FFC). 2001a. Capital Asset Management: Tools and Strategies for Decision-Making. Washington, D.C.: National Academy Press.

FFC. 2001b. Learning from Our Buildings: A State-of-the-Practice Summary of Post-Occupancy Evaluation. Washington, D.C.: National Academy Press.

FFC. 2001c. Sustainable Federal Facilities: A Guide to Integrating Value Engineering, Life-Cycle Costing, and Sustainable Development. Washington, D.C.: National Academy Press.

Federal Highway Administration (FHWA). 1999. Asset Management Primer. Washington, D.C.: Department of Transportation.

Gawande, Kishore, and Timothy Wheeler. 1999. Measures of effectiveness for governmental organizations. Management Science 45(1): 42-58.

General Accounting Office (GAO). 1996. Budget Issues: Budgeting for Federal Capital. GAO/AIMD-97-5. Washington, D.C.: GAO.

GAO. 1997. Defense Infrastructure: Demolition of Unneeded Buildings Can Help Avoid Operating Costs. NSIAD-97-125. Washington, D.C.: GAO.

GAO. 2000a. Park Service: Need to Address Management Problems That Plague the Concessions Program. GAO/RCED-00-70. Washington, D.C.: GAO.

GAO. 2000b. U.S. Infrastructure Funding Trends and Opportunities to Improve Investment Decisions. GAO-RCED/AIMD-00-35. Washington, D.C.: GAO.
GAO. 2001a. Federal Trust and Other Earmarked Funds: Answers to Frequently Asked Questions. GAO-01-199SP. Washington, D.C.: GAO.
GAO. 2001b. Human Capital: Meeting the Governmentwide High Risk Challenge. GAO-01-357T. Washington, D.C.: GAO.
GAO. 2001c. Major Management Challenges and Program Risks: A Governmentwide Perspective. GAO-01-241. Washington, D.C.: GAO.
GAO. 2001d. Public-Private Partnerships: Pilot Programs Needed to Demonstrate the Actual Benefits of Using Partnerships. GAO-01-906. Washington, D.C.: GAO.
GAO. 2002a. Contract Management: Interagency Contract Program Fees Need More Oversight. GAO-02-734. Washington, D.C.: GAO.
GAO. 2002b. Physical Infrastructure: Crosscutting Issues Planning Conference Report. GAO-02-139. Washington, D.C.: GAO.
GAO. 2002c. Water Infrastructure: Information on Financing, Capital Planning, and Privatization. Washington, D.C.: GAO.
GAO. 2003a. Alternative Approaches to Finance Federal Capital. GAO-03-1011. Washington, D.C.: GAO.
GAO. 2003b. Best Practices: Setting Requirements Differently Could Reduce Weapon Systems' Total Ownership Costs. GAO-03-57. Washington, D.C.: GAO.
GAO. 2003c. Budget Issues: Alternative Approaches to Finance Federal Capital. GAO-03-1011. Washington, D.C.: GAO.
GAO. 2003d. Creating a Clear Linkage Between Individual Performance and Organizational Success. GAO-03-488. Washington, D.C.: GAO.
GAO. 2003e. Federal Real Property: Vacant and Underutilized Properties at GSA, VA, and USPS. GAO-03-747. Washington, D.C.: GAO.
GAO. 2003f. High Risk Series: Federal Real Property. GAO-03-122. Washington, D.C.: GAO.
GAO. 2003g. Program Evaluation: An Evaluation Culture and Collaborative Partnerships Help Build Agency Capacity. GAO-03-454. Washington, D.C.: GAO.
GAO. 2003h. Tennessee Valley Authority: Information on Lease-Leaseback and Alternative Approaches to Finance Federal Capital. GAO-03-784. Washington, D.C.: GAO.
General Motors Corporation (GM). S.E.M. Facilities Portfolio Strategy. Materials provided to the committee in 2003.
General Services Administration (GSA). 1997. Governmentwide Review of Real Property Disposal Policy. Office of Governmentwide Policy. Washington, D.C.: GSA.
GSA. 1998. GPRIS Report, February. Washington, D.C.: GSA.
GSA. 1999. The Integrated Workplace: A Comprehensive Approach to Developing Workspace. Washington, D.C.: GSA.
Goldfein, Stephen M. 1994. The Base Realignment and Closure Commission: A Successful Strategy to Overcome Political Gridlock. Paper prepared for Roy Stafford, National War College.
Governmental Accounting Standards Board (GASB). 1994. Concepts Statement No. 2, Service Efforts and Accomplishments Reporting. Norwalk, Conn.: GASB.
GASB. 2003. Reporting Performance Information: Suggested Criteria for Effective Communication. Norwalk, Conn.: GASB.
Graham, J.R., and C.R. Harvey. 2001. The theory and practice of corporate finance: Evidence from the field. Journal of Financial Economics 60(2): 187-243.
Groppelli, A.A., and Ehsan Nikbakht. 2000. Finance, 4th ed. New York: Barron's Business Review Series.
Hamel, G., and C.K. Phahalad. 1994. Competing for the Future. Boston, Mass.: Harvard Business School Press.

Hammond, J.S., R.L. Keeney, and H. Raiffa. 1999. Smart Choices: A Practical Guide to Making Better Decisions. Boston, Mass.: Harvard Business School Press.

Hammond, K.R. 1996. Human Judgment and Social Policy: Irreducible Uncertainty, Inevitable Error, Unavoidable Injustice. New York: Oxford University Press.

Heerwagen, J.H. 2002. A balanced scorecard approach to post-occupancy evaluation: Using the tools of business. Learning from Our Buildings: A State of the Practice Summary of Post-Occupancy Evaluation. Washington, D.C.: National Academy Press.

Hudson, Ronald W., Ralph Haas, and Waheed Uddin. 1997. Infrastructure Management. New York: McGraw-Hill.

International Facility Management Association (IFMA). 1997. Views from the Top: Executives Evaluate the Facility Management Function. Research Report #17. Houston, Tex.: IFMA.

IFMA. 1998. Profiles '98. Houston, Tex.: IFMA.

IFMA. 2003. What Is Facility Management? Available at www.ifma.org. Accessed in January 2003.

Kadamus, D.A. 2003. Rescue your campus from stranded costs. Trusteeship (September/October): 28-32.

Kaplan, R.S., and David P. Norton. 1992. The balanced scorecard—measures that drive performance. Harvard Business Review (January-February): 71-79.

Kaplan, Robert, and David Norton. 2001. The Strategy-Focused Organization. Boston, Mass.: Harvard Business School Press.

Keeney, Ralph L., and Howard Raiffa. 1993. Decisions with Multiple Objectives. Cambridge, U.K.: Cambridge University Press.

Klein, Eva, William D. Middleton, and Tom Davies. 2002. Assessing Capital Equity. Working paper on the UNC program.

Kleindorfer, P.R., H.C. Kunreuther, and P.J.H. Schoemaker. 1993. Decision Sciences: An Integrative Approach. Cambridge, U.K: Cambridge University Press.

Lempert, R.J., S.W. Popper, and S.C. Bankes. 2003. Shaping the Next One Hundred Years. Santa Monica, Calif.: The RAND Pardee Center.

Linstone, H. 1984. Multiple Perspectives for Decision Making: Bridging the Gap Between Analysis and Action. New York, N.Y.: Elsevier-Science Publications.

Manning, Christopher A., and Stephen E. Roulac. 1996. Structuring the corporate real property function for greater 'bottom line' impact. Journal of Real Estate Research 12(3).

Mao, James C.T. 1970. Survey of capital budgeting: Theory and practice. Journal of Finance 25(2): 349-360.

McNeil, Sue. 2000. Asset management and asset valuation: The implications of the Government Accounting Standards Bureau (GASB) standards for reporting capital assets. Pp. 134-137 in Proceedings of the Mid-Continent Transportation Symposium 2000. Ames, Iowa: Center for Transportation Research and Education, Iowa State University.

National Academy of Public Administration (NAPA). 1996. Information Management Performance Measures: Developing Performance Measures and Management Controls for Migration Systems, Data Standards, and Process Improvements. Washington, D.C.: NAPA.

NAPA. 1998. Helpful Practices in Improving Government Performance. Washington, D.C.: NAPA.

National Council on Public Works Improvement (NCPWI). 1988. Fragile Foundations: A Report on America's Public Works. Washington, D.C.: U.S. Government Printing Office.

National Research Council (NRC). 1990. Committing to the Cost of Ownership: Maintenance and Repair of Public Buildings. Washington, D.C.: National Academy Press.

NRC. 1991. Pay Now or Pay Later: Controlling Cost of Ownership from Design Throughout the Service Life of Public Buildings. Washington, D.C.: National Academy Press.

NRC. 1996a. Understanding Risk: Informing Decisions in a Democratic Society. Washington, D.C.: National Academy Press.

NRC. 1996b. Affordable Cleanup? Opportunities for Cost Reduction in the Decontamination and Decommissioning of the Nation's Uranium Enrichment Facilities. Washington, D.C.: National Academy Press.
NRC. 1998. Stewardship of Federal Facilities: A Proactive Strategy for Managing the Nation's Public Assets. Washington, D.C.: National Academy Press.
NRC. 2000. Outsourcing Management Functions for the Acquisition of Federal Facilities. Washington, D.C.: National Academy Press.
NRC. 2003a. Completing the "Big Dig": Managing the Final Stages of Boston's Central Artery/Tunnel Project. Washington, D.C.: The National Academies Press.
NRC. 2003b. The Measure of STAR: Review of the U.S. Environmental Protection Agency's Science to Achieve Results (STAR) Research Grants Program. Washington, D.C.: The National Academies Press.
Noha, Edward A. 1993. Benchmarking: The search for best practices in corporate real estate. Journal of Real Estate Research 8(4): 511-523.
Nourse, Hugh O., and Stephen E. Roulac. 1993. Linking real estate decision to corporate strategy. Journal of Real Estate Research 8(4).
Office of the Inspector General (OIG), Department of the Treasury. 2002. General Management: The Mint Leased Excessive Space for Its Headquarters Operation. OIG-02-074. Washington, D.C.: OIG.
Office of Management and Budget (OMB). 1997. Capital Programming Guide: Supplement to Part 7 of Circular A-11. Washington, D.C.: OMB.
OMB. 2001. Federal investment spending and capital budgeting. Section 6 of Analytical Perspectives, Budget of the United States, Fiscal Year 2001. Washington, D.C.: OMB.
OMB. 2002. Analytical Perspectives, Budget of the United States, FY 2003. Washington, D.C.: U.S. Government Printing Office.
Okoroh, M.I., P.P. Gombera, and B.D. Ilozor. 2002. Managing FM (support services): Business risks in the healthcare sector. Facilities. 20(1-2): 41-51
Okoroh, M.I., C.M. Jones, and B.D. Ilozor. 2003. Adding value to constructed facilities: Facilities management hospitality case study. Journal of Performance of Constructed Facilities 17(1): 24-33.
O'Mara, Martha A. 1999. Strategy and Place: Managing Corporate Real Estate and Facilities for Competitive Advantage. New York: The Free Press.
O'Mara, Martha A, Eugene F. Page III, and Stephen F. Valenziano. 2002. The global corporate real estate function: Organisation, authority and influence. Journal of Corporate Real Estate 4(4): 334-347.
Pike, R.H. 1988. An empirical study of the adoption of sophisticated capital budgeting practices and decision making effectiveness. Accounting and Business Research (Autumn): 341-351.
Pint, E.M., and L.H. Baldwin. 1997. Strategic Sourcing: Theory and Evidence from Economics and Business Management. Prepared for the United States Air Force, Project Air Force. Santa Monica, Calif.: RAND.
Poterba, J.M. 1995. Capital budgets, borrowing rules, and state capital spending. Journal of Public Economics 56(2): 165-187.
President's Commission to Study Capital Budgeting (PCSCB). 1999. Report. Washington, D.C.: U.S. Government Printing Office. Available at clinton3.nara.gov/pcscb/report.html.
President's Task Force to Improve Health Care Delivery for Our Nation's Veterans. 2003. Final Report. Washington, D.C.: U.S. Government Printing Office.
PriceWaterhouse. 1993. A Guide to Public-Private Partnerships in Infrastructure: Bridging the Gap Between Infrastructure Needs and Public Resources.
Project Management Institute (PMI). 1996. A Guide to the Project Management Body of Knowledge. Sylva, N.C.: PMI Communications.
Roulac, Stephen E. 2001. Corporate property strategy is integral to corporate business strategy. Journal of Real Estate Research 22(1).

Schaffer, R.H., and Harvey A. Thomson. 1992. Successful change programs begin with results. Harvard Business Review (January/February): 80-89.
Schmidt, M.J. 2003a. Business Case Essentials: A Guide to Structure and Content. Available at www.solutionmatrix.com.
Schmidt, M.J. 2003b. What's a Business Case? And Other Frequently Asked Questions. Available at www.solutionmatrix.com.
Simons, Robert A. 1993. Public real estate management—Adapting corporate practice to the public sector: The experience in Cleveland, Ohio. Journal of Real Estate Research 8(4).
Sloan Program for the Construction Industry. 1998. Owner/Contractor Organizational Changes, Phase I Report. Austin, Tex.: University of Texas Press.
Slovic, P. 1993. Perceived risk, trust, and democracy. Risk Analysis 13(6): 675-682.
Stewart, J.D. 1984. The role of information in public accountability. Issues in Public Sector Accounting, Anthony Hopwood and Cyril R. Tomkins, eds. Oxford, U.K.: Phillip Allan.
Then, S.S. 1996. A study of organizational response to the management of operational property assets and facilities support services as a business resource—Real estate asset management. PhD thesis, Heriot-Watt University, Edinburgh, Scotland.
Then, D.S-S. 2003. Integrated resources management structures for facilities provision and management. Journal of Performance of Constructed Facilities 17(1): 34-42.
Tomkins, C. 1996. Strategic investment decisions: The importance of SCM. A comparative analysis of 51 case studies in U.K., U.S. and German companies. Management Accounting Research 7(2): 199-217.
Tracy, Bill. 2001. Space and asset management. Facility Design and Management Handbook, Eric Teicholz, ed. New York: McGraw-Hill.
U.S. Census Bureau. 2003. Annual Capital Expenditures: 2001. Washington, D.C.: U.S. Department of Commerce.
U.S. Department of Commerce (USDOC), Bureau of Economic Analysis. 2002. U.S. Statistical Abstract. Washington, D.C.: U.S. Government Printing Office.
U.S. Department of Transportation (USDOT), American Association of State Highway and Transportation Officials (AASHTO). 1996. Asset Management, Advancing the State of the Art into the 21st Century Through Public-Private Dialogue, FHWA-RD-97-046. Washington, D.C.: Federal Highway Administration.
USDOT. 1999. Asset Management Primer. Washington, D.C.: USDOT.
USDOT. 2000. Primer: GASB 34. Washington, D.C.: USDOT.
U.S. Department of Veterans Affairs (VA). 2003. Capital Investment Methodology Guide. Available at www.va.gov/budget/capital.
U.S. Government. 2002. The United States Government Manual, 2001-2002 Edition. Washington, D.C.: U.S. Government Printing Office.
U.S. Government. 2003. The U.S. Government's Official Web Portal, at HtmlResAnchor www.firstgov.gov/Agencies/Federal/Independent.shtml. Accessed June 12, 2003.
U.S. Navy Public Works Centers. Available at http://www.pwcwash.navy.mil. Accessed November 12, 2002.
Uzarski, D.R., D.K. Hicks, and J.A. Zahorak. 2002. Building and building component condition and capability metrics. Proceedings of the Seventh International Conference on Applications of Advanced Technology in Transportation, Kevin C.P. Wang, Samer Madanat, Shashi Nambisan, and Gary Spring, eds. Reston, Va.: American Society of Civil Engineers.
Varnier, D.J. 1999. Why Industry Needs Asset Management Tools. American Public Works Association (APWA), International Public Works Congress, NRCC Seminar Series "Innovations in Urban Infrastructure," pp 11-25. Kansas City, Mo.: APWA.
Weiss, C.H. 1998. Evaluation. 2nd Edition. Upper Saddle River, N.J.: Prentice Hall.
Whole Building Design Guide (WBDG). 2003. Available at www.wbdg.org. Accessed June 12, 2003.

Appendixes

A

Biographical Sketches of Committee Members

Albert A. Dorman, NAE, *Chair,* was the first president, chairman, and CEO (now retired) of AECOM Technology Corporation, one of the 200 largest private U.S. corporations. AECOM is a leading provider of program management and construction-related, diversified technical, professional services. The company maintains more than 100 offices worldwide and employs 17,000 people. Mr. Dorman has been involved in projects on all seven continents and has extensive experience in infrastructure programs, architecture, engineering, finance, and management in both the public and private sectors. In addition, he has served as chairman of a savings and loan association, has served on the boards of three publicly traded corporations, and has been a partner in more than two dozen real estate ventures. In the public arena, he has served as chairman of the Economic and Job Development Committee of the California Chamber of Commerce, served on the boards of five universities, and has delivered or published more than 30 papers. Mr. Dorman was elected to the National Academy of Engineering in 1998 for his contributions to the integration of civil engineering and architecture for large-scale public works projects. He is an honorary member of the American Society of Civil Engineers, also received its first award for Outstanding Lifetime Achievement in Leadership, and is a fellow of the American Institute of Architects. He holds a bachelor's degree in mechanical engineering and an honorary doctoral degree from the New Jersey Institute of Technology and a master's degree in civil engineering from the University of Southern California. He is also trained as an architect.

Adjo Amekudzi is an assistant professor in the School of Civil and Environmental Engineering at the Georgia Institute of Technology. Dr. Amekudzi's research focuses on managing infrastructure systems as assets. She specializes in applications of management methods, decision and risk analysis, and system performance evaluation. Her current research activities focus on asset valuation; environmental management systems; and applications of portfolio theory and sustainability metrics for infrastructure management. She has coauthored several papers. Dr. Amekudzi holds a B.S. in civil engineering from Stanford University, master's degrees in civil engineering and civil infrastructure systems from Florida International University and Carnegie Mellon University, respectively, and a Ph.D. in civil and environmental engineering from Carnegie Mellon University. She is a member of the American Society of Civil Engineers, the American Society for Engineering Education, the Transportation Research Board, the American Public Works Association, and the Georgia Transportation Institute.

Kimball J. Beasley is principal and manager of the New York office of Wiss, Janney, Elstner Associates, Inc. Mr. Beasley is a structural engineer with extensive experience in investigation and design of repairs for both historic and contemporary structures. During the past 30 years he has investigated more than 1,200 failures and performance problems involving a wide variety of building components and materials. His experience includes serviceability problems ranging from deterioration or water leakage and failure of traditional or composite wall systems to complete building collapse. He has served as a consultant on such notable buildings as the Metropolitan Museum of Art, Lincoln Center, the Whitney Museum of American Art, the World Wide Plaza, and the Empire State Building, in New York City; the City Hall in Richmond, Virginia; and the National Theater and Union Station in Washington, D.C. Mr. Beasley has authored over 30 articles and coauthored book chapters. He received a B.S. in structural and materials engineering from the University of Illinois and an M.B.A. from Pace University in New York.

Jeffery Campbell is a professor with the Facilities Management School of Technology, College of Engineering, at Brigham Young University (BYU). He has direct experience in and has published numerous articles on contract services, real estate, strategic market planning, project management, and life-cycle management. He has worked on research projects with the Association of Higher Education Facilities Officers, Dow Chemical, and the McKay School of Education at BYU. Prior to his position at BYU, Dr. Campbell held positions as an assistant professor in construction management and engineering at Boise State University; as director of construction and corporate facilities for Flying J, Inc.; as general/operations manager, FHP Medical Centers; as director of business development, Arnell-West General Contractors; and as a vice-president and commercial real estate broker with Smoot Commercial Brokers. Dr. Campbell has been honored by the International Facility Management Association with its Distinguished Educator Award. He received his Ph.D. from the University of Idaho.

Eric T. Dillinger is a principal and the vice president of facilities management for Carter & Burgess, Inc., consultants in planning, engineering, architecture, construction management, and related services. Mr. Dillinger has more than 14 years of experience in the area of facility management, including facility audits and condition assessment surveys, resource allocation, and capital asset management. He has participated in and directed facility audits and capital asset management programs for numerous federal government installations and agencies as well as private-sector organizations. Mr. Dillinger also has extensive experience in architectural and engineering endeavors, maintenance and repair prioritization, preventive and predictive maintenance, space utilization, inventory control, and scheduling and resource programming. He was a member of the National Research Council committee that authored the report *Stewardship of Federal Facilities: A Proactive Strategy for Managing the Nation's Public Assets* (1998). Mr. Dillinger has a B.S. in industrial engineering from Kansas State University and is a member of the International Facility Management Association, the Association of Higher Education Facility Officers, the Society of American Military Engineers, and the Society of American Military Comptrollers, among other organizations.

James R. Fountain, Jr., now retired, was assistant director of research at the Governmental Accounting Standards Board (GASB). The mission of GASB is to establish and improve standards of state and local governmental accounting and financial reporting that will result in useful information for users of financial reports and guide and educate the public, including issuers, auditors, and users of those financial reports. Mr. Fountain's current projects include financial condition reporting, capital asset use and infrastructure reporting, and concepts related to service efforts and accomplishments reporting. Prior to joining GASB, Mr. Fountain was director of finance for Fulton County, Georgia, and city auditor and later assistant city manager for Dallas, Texas. He is the coauthor of *Performance Auditing in Local Government* (GFOA, 1984), *Service Efforts and Accomplishments: Its Time Has Come* (GASB, 1990), *Elementary and Secondary Education* (GASB, 1989), *Report on the GASB Citizen Discussion Groups on Performance Reporting* (GASB, 2002), and *Reporting Performance Information: Suggested Criteria for Effective Communication* (GASB, 2003) and is the author of numerous articles. Mr. Fountain has a master's degree in public administration from Georgia State University and an M.B.A. in finance and a bachelor's degree in accounting from the University of Florida.

Thomas K. Fridstein is a managing principal with Hillier, a full-service architecture firm. Previously he was senior director of design at Tishman Speyer Properties, where he had worldwide responsibility for directing the design of major real estate development projects. Prior to joining Tishman Speyer Properties, Mr. Fridstein was a partner with the architectural firm Skidmore, Owings & Merrill.

He has more than 26 years of experience in the development, design, and construction of buildings, urban spaces, and interior environments in Europe, Asia, South America, and the United States. Mr. Fridstein was elected a fellow of the American Institute of Architects in 1996. He is a member of the Advisory Council of the College of Architecture, Art and Planning at Cornell University and the Board of Overseers of the College of Architecture at Illinois Institute of Technology and is active in the Council for Tall Buildings and Urban Habitat, the Urban Land Institute, and the American Institute of Architects. Mr. Fridstein holds a bachelor of architecture from Cornell University and an M.B.A. from Columbia University.

Lucia E. Garsys is the quality services executive officer for Hillsborough County, Florida, where she directs an organizational improvement initiative and performance audit function. Her experience in Fairfax County, Virginia, and Hillsborough County and her consulting experience in the Chicago metropolitan area include planning, development, redevelopment, capital investment using tax increment financing, public-private partnerships, and impact fees. In Hillsborough County, she directed initiatives to integrate operation and maintenance cost projections into capital budgeting, earmark revenues for building maintenance based on routine inventories, design an enterprise-wide system to manage a $600 million capital investment program, and develop project and contract management skills. She has consulted with the National Democratic Institute on planning and public participation in Lithuania and Belarus. Ms. Garsys is a member of the American Institute of Certified Planners. She holds a B.S. in city and regional planning from the Illinois Institute of Technology and a master's degree in urban planning from the University of Illinois at Urbana-Champaign.

David L. Hawk is a professor in the Schools of Management and Architecture at the New Jersey Institute of Technology (NJIT) and a visiting professor at the Helsinki University of Technology, Department of Industrial Engineering and Management. He has served as visiting professor/researcher at the Institute of International Business at the Stockholm School of Economics, Chalmers Technical University, and Tokyo Metropolitan University. He has also held a teaching position at Iowa State University. Dr. Hawk has published numerous articles and papers, lectured, and conducted extensive research on construction, building economics, and management issues. He is the author of *Foundation of a New Industry: Global Construction*. Dr. Hawk is a member of the International Trade and Finance Association, the International Society for Systems Sciences, and the American Association for the Advancement of Science and is a founding member of the European Academy of Management. He is the recipient of a number of honors, including the Year 2000 Robert W. Van Houten Award for Teaching Excellence at NJIT and the Progressive Architecture American Institute of Architects/Association of Collegiate Schools of Architecture Annual Research Award.

Dr. Hawk holds a bachelor of architectural engineering from Iowa State University, a master of architecture and master of city planning from the University of Pennsylvania, and a Ph.D. from the University of Pennsylvania, Wharton School of Business. He does extensive consulting in the fields of relationship alignment and governance systems.

Ralph L. Keeney, NAE, is a research professor in decision sciences at the Fuqua School of Business of Duke University. He previously taught at the Marshall School of Business and the Department of Industrial and Engineering Systems at the University of Southern California. Dr. Keeney has been a consultant for numerous public and private organizations, working in the areas of large-scale siting studies, energy policy, environmental and risk studies, and corporate management problems. He has been a professor of engineering and management at the Massachusetts Institute of Technology, a research scholar at the International Institute for Applied Systems Analysis in Austria, and founder of the decision and risk analysis group of a geotechnical and environmental consulting firm. Dr. Keeney was elected to the National Academy of Engineering in 1995 for contributions to the theory and engineering practice of decision analysis as applied to complex public problems with conflicting objectives. Dr. Keeney has a Ph.D. from MIT and did his undergraduate work in engineering at the University of California at Los Angeles.

Stephen J. Lukasik is a consultant to SAIC and a former visiting professor at the Georgia Institute of Technology. Dr. Lukasik has also been a visiting scholar at the Stanford University Center for International Security and Cooperation, where his research focused on technical and policy issues related to critical infrastructure protection. Dr. Lukasik is a former director of the Defense Advanced Research Projects Agency and a former chief scientist of the Federal Communications Commission. In addition, he has held various senior positions in industry, including vice president of TRW, Inc., the Xerox Corporation, the Northrop Corporation, and RAND Corporation. Dr. Lukasik has served on a number of National Research Council committees, including the Committee on Scientists and Engineers in the Federal Government. He is a member of the American Association for the Advancement of Science, the American Physical Society, Sigma Xi, and Tau Beta Pi. Dr. Lukasik received his B.S. in physics from Rensselaer Polytechnic Institute and an M.S. and a Ph.D. in physics from the Massachusetts Institute of Technology.

RADM David Nash, U.S. Navy Civil Engineers Corps (ret.), *Vice Chair*, is currently serving as Director of the Iraq Infrastructure Reconstruction Office. He is also the vice president of government operations at BE&K, Incorporated, a privately held international design-build firm that provides engineering, construction, and maintenance for process-oriented industries and commercial real estate projects. Prior to joining BE&K, Admiral Nash was president of PB Buildings, Inc., and was formerly manager of the Automotive Division of Parsons Brinckerhoff Construction Services (PBCS), Inc. Admiral Nash served for 33 years in the U.S. Navy, completing his career as commander of the Naval Facilities Engineering Command and chief of civil engineers of the U.S. Navy. He is a member of the Society of American Military Engineers, the American Society of Civil Engineers, the American Public Works Association, the National Society of Professional Engineers, the Institute of Electrical and Electronics Engineers, and the American Society of Quality Control. He holds a B.S. degree in electrical engineering from the Indiana Institute of Technology and an M.S. in financial management from the Naval Post Graduate School in Monterey, California. Admiral Nash is a registered professional engineer in Pennsylvania and Michigan.

Carol Ó'Cléireacáin is a nonresident senior fellow at the Brookings Institution's Center for Urban and Metropolitan Policy and an independent economic and management consultant in New York City. Dr. Ó'Cléireacáin was a member of the President's Commission to Study Capital Budgeting and also served on the National Civil Aviation Commission chaired by Norman Mineta. She was a nonresident senior fellow and visiting fellow at the Brookings Institution; budget director and finance commissioner of the City of New York under Mayor David Dinkins; senior research associate at the Bildner Center, CUNY Graduate Center; chief economist at District Council 37 AFSCME in New York City; and held a number of adjunct teaching positions, including at Barnard College, Columbia University, the Wagner Graduate School at NYU, and the Milano Graduate School at the New School University. She currently serves on two corporate boards—as director and chair of the audit committee of Spectrum Pharmaceuticals and as director and member of the Executive Committee of Trillium Asset Management. She is the author/editor of several books and numerous articles and papers on budgeting and management. Dr. Ó'Cléireacáin holds a Ph.D. in economics from the London School of Economics and an M.A. and a B.A. in economics from the University of Michigan.

Charles Spruill is the manager for space, project, and facilities services at Fannie Mae. Prior to joining Fannie Mae, he was a facility management professional with Marriott International. He has 18 years of progressive experience encompassing space management, project management, asset and inventory management, operations management, and lease and property management. As the facility manager of Marriott International Headquarters in Bethesda, Maryland, Mr. Spruill administered 1 million square feet of office, amenity, and support space for 3,500 associates. He was responsible for policies and procedures related to office space and project delivery and for initiating and implementing recommendations for cost savings and improvements in operational effectiveness. He maintained the corporate strategic facility forecast, consisting of immediate, interim, and long-term space needs, and cash flow and budget impact analysis. In addition, Mr. Spruill was instrumental in establishing internal cost control and recovery mechanisms, office space standards, project documentation standards, and contract administration procedures. Mr. Spruill received a B.F.A. in interior design from Virginia Commonwealth University.

B

Committee Interviews and Briefings

2002

January 29 First Committee Meeting; briefings by William Brubaker, Director, Facilities and Engineering Operations, Smithsonian Institution, and Captain Patrick Layne, Chief, Office of Civil Engineering, U.S. Coast Guard

April 9 Second Committee Meeting; briefing by Craig Crutchfield, Program Examiner, Office of Management and Budget

June 6 Interview with Wendy Comes, Executive Director, Federal Accounting Standards Advisory Board

June 13 Interview with Jeanne Wilson, Republican Staff Assistant, House Appropriations Committee, Subcommittee on Energy and Water Development

June 20 Interview with William Fife, Corporate Vice President, DMJM + Harris

June 21 Interview with Rusty Hodapp, Director, Airport Engineering and Maintenance, Dallas/Fort Worth International Airport, and Jack Allison, Manager of Infrastructure Maintenance

APPENDIX B

June 27	Interview with James O'Keeffe, Senior Transportation Analyst, Senate Budget Committee
June 28	Interview with David Skiven, Executive Director, Worldwide Facilities Group, General Motors Corporation
June 28	Interview with Gen. Charles Williams, Director/Chief Operating Officer, Bureau of Overseas Buildings Operations, U.S. Department of State
August 14	Interview with Harry Olsen, Project Director, LCOR, Inc.
August 15	Interview with Rudy Umscheid, Vice President of Facilities, USPS; Diane Van Loozen, Director, Facilities Programs; Michael Goodwin, Director, Design and Construction; Mike Mattera, Manager, Facilities Planning and Approval
August 22	Interview with Benjamin Montoya, former head of Public Utilities Service Company of New Mexico
September 9	Follow-up interview with William Fife, Corporate Vice President, DMJM + Harris
September 17	Interview with Thomas Kowalyk, Johnson and Johnson Corporation
October 7	Interview with Thomas D. Farrell, Managing Director, and Katherine Farley, Senior Managing Director, Tishman Speyer Properties
November 15	Interview with representatives of Intel Corporation
November 15	Interview with Dennis Cuneo, Senior Vice President, Toyota Motors North America, Inc.
November 21	Interview with Douglas Hansen, Director, Installations Requirements and Management, Installations and Environment, Undersecretary for Acquisition, Technology, and Logistics, Department of Defense

2003

January 7   Informational meeting with staff of the General Accounting Office, chaired by Bernard Ungar

January 7   Informational meeting with staff of the Department of the Navy, chaired by RADM Michael Johnson, Naval Facilities Engineering Command

C

Interview Discussion Outline

For purposes of this interview, facilities investment includes new construction, renewal, maintenance, retrofitting, replacing and decommissioning of facilities.

1. How would you characterize your organization's role in making decisions about facilities investment?
   - ❑ Own facilities
   - ❑ Lease facilities
   - ❑ Provide facilities to others
   - ❑ Use facilities
   - ❑ Manage facilities
   - ❑ Approve facility projects
   - ❑ Approve funding for facility projects
   - ❑ Track/audit expenditures for facilities projects
   - ❑ Measure performance of facility projects
   - ❑ Other

2. Does your organization have an inventory of facilities and their condition?
3. What is the mission of your facilities investment/management organization?
4. How are your organization's goals and objectives integrated into the decision-making process for facility investment?
5. How does your organization document objectives to be satisfied by facility investment?

6. What performance measures/metrics are used to evaluate the results of facility investments? (For example, rate of return, discounted cash flow, nonfinancial indicators). At what point in the process are they used? At what level of functionality are they applied (project level, portfolio level, both)?
7. How does your organization define success for facility investments? Define failure/inadequate performance for facility investments?
8. How does your organization identify facilities projects/opportunities for facilities investment (market analysis, specific search based on strategic goals, conduct a comprehensive needs assessment, gap analysis between current and needed capabilities, other)?
9. How does your organization document the need for facilities?
10. How does your organization identify and evaluate alternative approaches for facility investment (build new, lease, purchase, renew/retrofit existing)? Who is involved and what criteria are used for establishing alternatives?
11. How does your organization quantify the costs, benefits, risks, and trade-offs of alternatives?
12. How does your organization rank and select projects?
13. How does your organization make trade-offs among facility projects and other organizational objectives/programs?
14. How does your organization obtain funding for facilities?
15. Does your organization use a top-down or bottom-up approach to fund facility investments?
16. Who must approve your facilities investment budget/revenue and operating plan internally? Externally? (name of groups, positions, not persons)
17. What type of innovative approaches to full up-front funding are considered? How do you weigh alternatives to full up-front funding?
18. How is your organization's long-term capital plan prioritized for the current operating year?
19. How do you estimate the availability of funding? How do you know how much money you have available to spend?
20. When acquiring or retrofitting a facility, does your organization have a long-term expectation for the use of the space? Design flexibility into the facility to accommodate unexpected or multiple uses? Conduct a life-cycle cost analysis?
21. Does your organization develop an up-front exit strategy for a facility investments—that is, a plan for getting out of a facility investment at any time at a reasonable cost?
22. How does your organization approach decisions related to operating and maintaining facilities?
23. How does your organization decide that money should be invested to renew or retrofit a facility?
24. How does your organization decide to decommission a facility? Who is involved in the decision process? What criteria are used?

25. Who is responsible for reviewing projects after a fixed period of usage to determine whether the alternative assumptions were correct?
26. How does your organization incorporate lessons learned into the decision-making process?
27. Please share any other comments/information that you believe may be of value to the committee for its study.